Deutsches Zentrum für Entwicklungstechnologien – GATE

Deutsches Zentrum für Entwicklungstechnologien – GATE – stands for German Appropriate Technology Exchange. It was founded in 1978 as a special division of the Deutsche Gesellschaft für Technische Zusammenarbeit (GTZ) GmbH. GATE is a centre for the dissemination and promotion of appropriate technologies for developing countries. GATE defines „Appropriate technologies" as those which are suitable and acceptable in the light of economic, social and cultural criteria. They should contribute to socio-economic development whilst ensuring optimal utilization of resources and minimal detriment to the environment. Depending on the case at hand a traditional, intermediate or highly-developed can be the „appropriate" one. GATE focusses its work on the key areas:

– *Dissemination of Appropriate Technologies:* Collecting, processing and disseminating information on technologies appropriate to the needs of the developing countries; ascertaining the technological requirements of Third World countries; support in the form of personnel, material and equipment to promote the development and adaptation of technologies for developing countries.

– *Environmental Protection:* The growing importance of ecology and environmental protection require better coordination and harmonization of projects. In order to tackle these tasks more effectively, a coordination center was set up within GATE in 1985.

GATE has entered into cooperation agreements with a number of technology centres in Third World countries.

GATE offers a free information service on appropriate technologies for all public and private development institutions in developing countries, dealing with the development, adaptation, introduction and application of technologies.

Deutsche Gesellschaft für Technische Zusammenarbeit (GTZ) GmbH

The government-owned GTZ operates in the field of Technical Cooperation. 2200 German experts are working together with partners from about 100 countries of Africa, Asia and Latin America in projects covering practically every sector of agriculture, forestry, economic development, social services and institutional and material infrastructure. – The GTZ is commissioned to do this work both by the Government of the Federal Republic of Germany and by other government or semi-government authorities.

The GTZ activities encompass:

– appraisal, technical planning, control and supervision of technical cooperation projects commissioned by the Government of the Federal Republic or by other authorities

– providing an advisory service to other agencies also working on development projects

– the recruitment, selection, briefing, assignment, administration of expert personnel and their welfare and technical backstopping during their period of assignment

– provision of materials and equipment for projects, planning work, selection, purchasing and shipment to the developing countries

– management of all financial obligations to the partner-country.

Deutsches Zentrum für Entwicklungstechnologien – GATE
in: Deutsche Gesellschaft für Technische Zusammenarbeit (GTZ) GmbH
P. O. Box 5180
D-6236 Eschborn
Federal Republic of Germany
Telephon: (06196) 79-0 Telex: 41523-0 gtz d Fax: (06196) 794820

Henrik Norsker, James Danisch

Glazes – for the Self-Reliant Potter

A Publication of Deutsches Zentrum für Entwicklungstechnologien – GATE
A division of the Deutsche Gesellschaft für Technische Zusammenarbeit (GTZ) GmbH

vieweg

The authors:
Henrik Norsker has been making pottery since 1970. He left his pottery workshop in Denmark in 1976 to establish a pottery school in a village in Tanzania. Since then he has continued working in developing countries with promotion of small scale ceramics industries. Besides Tanzania he has been employed in ceramics projects in Burma, Bangladesh and Nepal.

James Danisch has been making, selling and experimenting with ceramics since 1963. He has taught college level ceramics in Scotland and California, and has conducted workshops in the US, South America and Canada. From 1984 to 1992, he has been working with small scale and rural ceramics develolpment in Nepal. His articles on ceramics have been published in several magazines, and he has studied traditional and modern techniques in Europe, Nepal, India, Thailand, Burma, South America and Mexico.

Books by the same author include:
The Self-Reliant Potter: Refractories and Kilns
GATE / Vieweg, 1987

Clay Materials – for the Self-Reliant Potter
GATE / Vieweg, 1990

Forming Techniques – for the Self-Reliant Potter
GATE / Vieweg, 1991

Die Deutsche Bibliothek – CIP-Einheitsaufnahme

Norsker, Henrik:
Glazes : for the self-reliant potter ; a publication of Deutsches
Zentrum für Entwicklungstechnologien GATE,
a division of the Deutsche Gesellschaft für Technische Zusammenarbeit
(GTZ) GmbH / Henrik Norsker ; James Danisch. –
Braunschweig : Vieweg, 1993
ISBN 978-3-528-02067-5 ISBN 978-3-663-06865-5 (eBook)
DOI 10.1007/978-3-663-06865-5
NE: Danisch, James:

The author's opinion does not necessarily reprensent the view of the publisher.

Published by Friedr. Vieweg & Sohn Verlagsgesellschaft mbH, Braunschweig
Vieweg is a subsidiariy company of the Bertelsmann Publishing Group International.

ISBN 978-3-528-02067-5

CONTENTS

1. INTRODUCTION AND SCOPE

There are already many books in the world on the subject of ceramic glazes. So the obvious question is: why yet another book on the subject? The authors have worked together for several years in a ceramics development project in Nepal, which is based on using local raw materials and resources. There are few existing books which offer much help in this area, especially working in the low temperature range from 900°C-1100°C, where lead glazes have been the tradition but which now, with greater understanding of health hazards, need to be replaced with lead-free glazes. This book is intended to provide practical information for ceramists working in developing countries, with little access to the prepared and controlled glaze materials available in industrialized nations.

FIGURE 1.0.0.A Engobe decorations with transparent glaze. Nepal.

Glazes are at one and the same time the area of most fascination and most difficulty for potters. Most potters have little inclination or time to devote to developing glazes, faced as they are by the daily need to produce for the market. However, there often are times when familiar glazes suddenly stop working correctly or special glazes are required for customers. This book offers guidelines for developing and altering glazes, understanding where problems with glazes come from, and standard procedures for testing and developing glazes when there is no laboratory equipment available. It has been written for potters who have little knowledge of chemistry and mathematics.

1.1. GLAZE MAKING USING LOCAL MATERIALS AS FAR AS POSSIBLE

Most small producers of glazed ceramics will use glazes that are prepared by a company specializing in supplying industry. However, these glazes are often unreliable, as big companies tend to serve large-scale producers and have little interest in the special glazes needed by small industries. For that reason, the small producer is often forced to rely on his own glaze production, with little or no laboratory equipment available. Additionally, the small producer does not usually have access to raw materials at reasonable prices, so he must use locally available raw materials that do not have an accurate chemical analysis.

1.2. GLAZE AND CLAY SYSTEMS

The producer must think carefully before starting production. When a particular glaze is wanted, it must work with the available clay body, production system and firing system. For example, if you only have low-temperature red clay available, your glazes must work at around 1050°C. If you only have coal available for firing, you must make sure that it will work for your product. The following chapters provide information which will help you to make these decisions.

2. THE NATURE OF GLAZES

2.1. GLASS AND GLAZE, BENEFIT OF GLAZE

Glass is a useful material that has been known for thousands of years. It can be produced in many different shapes for many purposes, and it has many useful qualities: it is transparent, hard, resistant to chemicals, and can have many colors.

Glaze is a special type of glass, made for coating ceramic products. Whereas glass is suitable for forming into bottles or windows, glaze is different because it is applied on a ceramic surface and must form a hard, durable coating after being melted in the kiln. It must not run off the product and must stay on the product after firing without cracking.

2.2. GLAZE MAKING IS DIFFICULT

Because glaze before firing looks nothing like the finished product and because we are not able to directly understand what happens when glaze melts at high temperatures, making glazes is very difficult. We must try to understand which materials melt at certain temperatures and what happens when materials are combined. It requires a lot of direct experience before you start to understand causes and effects. In this way it is like cooking: we are familiar with cooking because we know the raw materials, and by trial and error we have a good idea of what the finished meal will be like. However, imagine that you are in a foreign country with unfamiliar food in the market and you want to make a meal: How do you start? The best way is with a cookbook full of recipes and a local friend to tell you if the result is correct or not. This book is intended as a cookbook for the independent potter.

Although by just reading this book and experimenting with it you will probably be able to make glazes after some time, there is no substitute for learning about glazes from an experienced teacher, who can save you a lot of time by guiding you in proven directions.

2.3. HISTORY OF GLAZES

Unglazed ceramics have been in existence for over 10,000 years. It has only been in the last 2000 years that there have been glazed ceramics and only in the last 100 years that a scientific approach to glaze making was developed. For that reason, glazes still occupy a mysterious area somewhere between science and magic.

The first glazes were probably invented in

middle eastern countries, where there naturally exist deposits of sodium and potassium compounds (soda ash and pearl ash) that melt at low temperatures (800°-1000^oC). By chance, early potters discovered that some clays when put in the fire developed a shiny surface. These self-glazing clays are known as "Egyptian paste". They are not very useful for making household items, being difficult to form.

The next step was to develop these substances so that they could be applied to the surface of pottery clay in order to give it the desirable qualities of a hard, shiny, easy-to-clean and durable surface. Because early potters did not have the technology to reach high firing temperatures, they had to use

FIGURE 2.3.0.A Decoration with a red ochre clay without the use of glaze. Ban Chiang, Thailand.

materials with low melting points, mainly sodium, potassium and lead compounds. Glaze development had to be done by trial and error, since these early potters had no idea of chemistry. This took a lot of time and effort, and naturally successful glazes were closely guarded secrets. These early glazes were often soft and not durable, and had problems such as cracking and eventually falling off the pot. Additionally, glazes based on lead were poisonous both for the potters who worked with them and for users.

It was only when potters learned to reach high temperatures that truly permanent ceramics were developed. There are many more common chemicals and minerals that melt above 1100^oC to form glazes, and clay that is fired to these high temperatures is also much stronger and resistant to water.

2.4. CLASSIFICATION (EARTHENWARE, STONEWARE ETC.)

Although there are many different ways to classify glazes, the simplest way to understand them is according to the firing temperature. The useful range of temperatures for glaze melting is from 900°-1300^oC. In this book, we talk about two different categories of glaze:

- low temperature from 900-1100^oC, called earthenware;
- high temperature from 1100-1300^oC, called stoneware.

These two categories are used because they require different raw materials as the main ingredients of the glaze.

3. TEMPERATURE RANGES AND REQUIREMENTS

3.1. WHAT IS TEMPERATURE?

Temperature means the amount of heat energy in a material. We raise the temperature of a material by providing it with heat energy, using a fire or electricity. What effect does this have on a material? We know that many familiar substances can exist in different states of solid, liquid and gas. For example, water can exist as ice, liquid water or steam. What is different about it? Only the temperature. All materials consist of atoms and molecules which are in constant motion. The amount of motion depends on the temperature. Cold materials have less motion and therefore appear solid to us (e.g. ice). When the temperature is increased, the motion of the molecules becomes greater and they can move more freely around each other (e.g. water). When the temperature is increased even more, the molecules become very active, as we can see when water boils. Then the molecules are even less bonded together and we see gas (e.g. steam).

Similarly, glazes are solid when they are cold (at room temperature), liquid when they are heated sufficiently (in the kiln), and become gas when they are heated too much.

It is also important to understand the relationship between clay and glaze. Most common red clay (such as brick clay) melts by 1100°C. This makes it useful for forming low temperature products. Above 1200°C, it can be used as a glaze.

3.2. LOW TEMPERATURE RANGE 900-1100°C

Products called earthenware, whiteware, low-temperature ceramics, and terra cotta are all fired in the range of 900-1100°C. We will call these products generally "earthenware". What they have in common are clay bodies that develop their maximum strength in this range, and glazes that are based on low-melting compounds such as lead, sodium and potassium.

3.2.1. ADVANTAGES/ DISADVANTAGES

Advantages: low temperature ceramics have the advantage of easy firing – it is much simpler to construct kilns and burner systems that have to reach no more than 1100°C, and fuel costs are lower. Bright colors are possible in this range. Most common clays cannot be fired higher than this.

Disadvantages: earthenware is often not as strong as high temperature ware, because the clay does not become vitreous. This means that it also has some porosity (the property of absorbing water) with the result that earthenware products often do not hold water unless the glaze is perfectly fitted to the body. Also, it is easier to chip the glaze away from the clay.

Historically, many earthenware glazes were based on poisonous lead because it is easy to melt: nowadays this is not a problem because lead can be replaced by non-poisonous materials.

Modern earthenware glazes are usually based on frits, which are expensive – the lower firing cost must be compared to the higher cost of the glaze.

3.2.2. APPROPRIATE PRODUCTS

Earthenware is used for all common household containers – cups, bowls, storage containers, oil lamps etc. Ordinary wall tiles, most low-cost tableware, sanitary ware, common unglazed containers, bricks, roof tiles etc. are all made in the low temperature range. Many countries have a long tradition of glazed red clay products, which are still useful in modern times. Most modern factories have changed their production to white clay products, which have become more feasible with recent developments of white bodies that become strong enough at low temperatures.

3.2.3. CLAY/GLAZE CHARACTERISTICS

Earthenware clay

Common red-burning clay is normally used, often mixed with talcum powder to increase its firing range. In many countries, red clay which contains lime is used because it makes it easier to formulate glazes that do not craze (crack). White firing clay bodies are often based on talc, ball clay and fluxes to make them harder.

FIGURE 3.2.3.A Earthenware lion made by traditional potters is now produced with a transparent glaze. Bhaktapur, Nepal.

Earthenware glaze

Earthenware glazes are based on low-melting materials, mainly lead oxide (white lead oxide, red lead oxide), sodium and boron compounds (soda ash, borax, boric acid) and potassium compounds (pearl ash, also known as potassium carbonate). Usually it is necessary to use these compounds in the form of frits (see chapter on frits).

3.2.4. RAW MATERIAL REQUIREMENTS

Most of the raw materials for low temperature glazes can be obtained from commonly available sources. They include: local clays, wood and rice husk ash, limestone, and even soap powder (based on sodium and boron compounds). Materials such as borax must be obtained from chemical suppliers. Ready-made frits can be obtained from glaze suppliers, but in many locations it is necessary to make them from raw materials.

3.3. HIGH TEMPERATURE RANGE 1100-1300°C

Types of ware fired in this range are known as stoneware and porcelain.

3.3.1. ADVANTAGES/ DISADVANTAGES

Advantages

High temperature products are generally stronger, more acid and abrasion-resistant. Raw materials do not require fritting. The clay is more vitreous and thus does not have problems of water seepage.

Disadvantages

Kilns for high temperatures require more sophisticated bricks and kiln furniture, and better burner systems. Fuel costs are higher.

3.3.2. APPROPRIATE PRODUCTS

High temperature products include stoneware utilitarian items, whiteware of various types, porcelain and electrical insulators.

3.3.3. CLAY/GLAZE CHARACTERISTICS

Clay

Clay body raw materials are limited to those clays which can withstand high temperatures without melting: fireclays, ball clays, china clays, "stoneware" clays. Most bodies also include feldspar to cause vitrification, which prevents water seepage through the body.

Glaze

High temperature glaze is easier to make than the low temperature sort, mainly because it is not necessary to frit the ingredients.

3.3.4. RAW MATERIAL REQUIREMENTS

Most stoneware and porcelain glazes are based on feldspar, quartz, limestone and clay, with other ingredients to provide specific properties of surface, color etc.

3.4. FIRING SYSTEMS AND GLAZE EFFECTS

Different types of kilns and fuels have specific effects on glaze color and surface.

3.4.1. OIL, GAS, WOOD, COAL, ELECTRICITY, OTHER

These are the main options for fuel. Each fuel requires a different kiln design and burner system. You must first decide which fuel is most available and most economical. The

FIGURE 3.4.1. A Glazed ware set in saggars.

choice of fuel will determine whether products can be open-fired on shelves, or whether it is necessary to use saggars to protect the glaze from ash and contamination from dirty fuel.

The cost of fuel should be thought about very carefully. One kg of fuel produces a certain amount of heat. Heat is usually measured in *calories* or in *British Thermal Units (BTU)*. One calorie is the amount of heat required to raise the temperature of one cubic centimeter of water 1°C. The table at page 170 shows the heat value of different fuels. Because a calorie is very small, the usual unit of heat is expressed as *kilocalories* (kilo = 1000, so 1 kilocalorie = 1000 calories).

A particular kiln, loaded with an average number of products and fired to a specific temperature, will usually require the same amount of fuel each time, since it requires a specific number of calories to convert raw clay and glaze into finished ceramics. When you know the total kg of products and the total cost of one firing, it is easy to calculate the cost per kg of product:

$$\frac{\text{Total cost}}{\text{Total kg}} = \text{cost per kg}$$

You can also calculate the total number of calories required to do one firing. If you are using kerosene, you can find from the table that one liter of kerosene supplies about 12,000 kilocalories of heat. So, if you use 80 liters to do a firing, the calculation is:

(Total fuel) X (kilocalories per unit) = total kilocalories required

80 X 12,000 = 960,000 kilocalories

When deciding on the type of fuel to use, you should find out the cost per kilocalorie for different fuels in your area.

Oil

Oil is available in many different forms, all of which can be used by the potter, including kerosene, diesel, furnace oil, and waste crankcase oil. Kerosene is the most clean-burning (without too much smoke or impurities), and waste crankcase oil is the dirtiest to use. Normally, products can be open-fired, but oil will produce some discoloration. For high quality whiteware, saggars may be necessary. Oil is suitable for high or low temperatures.

Oil provides between 9000 and 11000 kilocalories per kg.

Gas

Gas is available as natural gas, producer gas or liquid propane gas. Where gas is available at a reasonable cost (compared to other fuels), it is the easiest fuel to use. Gas is very clean-burning, does not require sag-

FIGURE 3.4.1.B Open setting in an oil-fired kiln.

gars, and the burners are also simple to manufacture locally. It is suitable for any temperature.

Wood

Almost any kind of wood can be used for firing kilns. Nowadays, wood is a scarce resource in most countries and more and more it is being replaced by other fuels. Firing with wood is labor-intensive. Because it produces a large volume of ash, it is usually necessary to fire the ware in saggars. It is suitable for any temperature.

On the other hand, wood is a renewable resource and in many areas of the world it is produced as a cash crop, which makes it appropriate to use.

The calorific value of wood is difficult to calculate, because it depends on the type of wood, whether it is wet or dry, and the efficiency of burning. Dry wood can supply between 3000 and 4500 kilocalories per kg, whereas the same wood when wet may produce only half the calories.

Coal

Coal comes in many different grades, all of which are suitable for firing kilns. Firing with coal is labor-intensive, but in many countries it is the cheapest fuel available. Coal also produces ash and impurities, so it is usually necessary to fire the ware in saggars. It is best for high temperatures, but can be used at any temperature.

Coal can provide between 4500 and 7700 kilocalories per kg.

Electricity

Electric kilns are practical for the small producer where there is a reliable source of electricity. Because there is no combustion, electricity is the cleanest fuel of all. Electric kilns fire very evenly and do not require saggars. Electricity is best for temperatures up to 1100°C.

Other fuels

These include tires, which burn very well but produce a lot of smoke, and also produce poisonous gases. They can be used in kilns designed to burn wood or coal. Some brick industries use scrap asphalt from roads as fuel. Also in this category are such fuels as brushwood, sawdust and rice husk. Most of these are dirty-burning, so require the use of saggars. They are best for low temperatures.

3.4.2. OXIDATION/REDUCTION

To understand oxidation and reduction, it is necessary to know how fuel burns. All fuel produces heat when it combines with the oxygen in the air. As anyone knows who has made a wood fire, if there is plenty of air the fire burns hot and clean, with little smoke. This is called an oxidation fire. If the air is reduced, there will be less heat and more smoke. This is called a reduction (or reducing) fire, which simply means reducing the amount of oxygen. So:

- Oxidation firing means there is plenty of air and no smoke.
- Reduction firing means there is little air and more smoke.

Glazes will have different colors and surfaces depending on whether they are fired in oxidation or reduction conditions. Oxidation has its greatest effect on the metallic oxides that are used to create color in glazes. For example:

OXIDE	OXIDATION	REDUCTION
red iron oxide	brown	red-brown, black
copper oxide	green, blue	red

Iron also changes from a gray color to a red color when it rusts. This is because oxide from the air combines with the metal and forms iron oxide.

In firing, it is difficult to exactly control the amount of oxidation or reduction. Many beautiful glazes can be obtained in reduction firing, so it is widely used for decorative stoneware, and for lusterware. However, the results are variable and difficult to reproduce every time, and even in one kiln-load there will be differences. For that reason, most producers who need to supply a uniform product use oxidation firing.

3.4.3. VAPOR GLAZING

In vapor glazing techniques, the glaze is not applied to the product before firing in the usual manner. Instead, glaze is introduced into the kiln through the firebox at the end

FIGURE 3.4.3.A A traditional salt-glazed jug made with a cobalt blue decoration. Salt-glazed stoneware has been made in the Rhineland since the 14th century.

of the firing, when there is enough heat to change the glaze into vapor form. The most common material for vapor glazing is ordinary salt. At temperatures above 1100°C, salt breaks down into sodium and chlorine vapor, which circulates through the kiln. The sodium is attracted to silica in the clay and forms a strong, durable glaze. Salt glazing is used mainly for sewage pipes, because it is cheap and a perfectly glazed surface is not necessary. In Europe, it was once used widely for household items, even including beer bottles. Nowadays, salt glazing is less popular because it produces toxic smoke that harms the environment.

Salt is sometimes replaced by soda ash and sodium bicarbonate, which produce a similar vapor glaze without the poisonous side effects. Vapor glazing is not recommended for the small producer, except for making specialized art ceramics.

4. DECISIONS

As a ceramics entrepreneur, you must start by making decisions: what product? what temperature? how much technology? These decisions depend on your market, raw material and fuel availability. In industrialized countries, where everything is easily available, the decision will usually be based first on the market, and then the best combination of clay body, glazes and kiln can be decided on.

In developing countries, it is usually necessary to start by thinking about raw materials and fuel. Then the product can be selected.

Usually, it is easiest to use the same technology as other producers, as most of the problems will have already been solved. On the other hand, a new type of technology can capture a new market sector with no competition. However, a new technology may cause technical problems that a potter cannot solve without outside help.

Some typical questions for the entrepreneur to answer are given below.

4.1. SELECTING YOUR BEST FIRING TEMPERATURE

Is high-firing clay available?

If so, it may be best to decide on high temperature ceramics (stoneware or porcelain), as producing reliable glazes will be easier.

Is only low-firing clay available?

If so, it will be necessary to select a low temperature system and to make frits or purchase ready-made ones.

Are ready-made glazes available?

If there is a reliable source of glazes nearby, a lot of trouble can be saved by using these.

What are the fuel constraints?

If only electric firing is available, then only low temperature systems will be practical. If oil or coal is available, the additional costs of using saggars should be compared with the cost of clean-burning fuels.

4.2. MARKET FACTORS

Most producers decide to enter the ceramics sector because there is already a good market and not enough local supply or because they think they can create a market for products that are not yet common in their area.

What is the existing market?

For example, if there is already a good market for glazed white earthenware (perhaps imported), the potential producer will have to find out if he can produce similar products at competitive prices. If he wants to compete directly, he will have to take up the same clay/glaze/firing system.

FIGURE 4.2.0.A In Germany stoneware bottles are still produced for liquor.

FIGURE 4.2.0.B In Nepal locally made sewage pipes replace imported pipes.

Is there a possible new market?

On the other hand, it may be possible to produce a product with the same function, but using a less costly technology. For example, it may be possible to produce glazed red clay earthenware cheaper than the whiteware on the market and thus to create a new market.

Small-scale vs. large-scale

Large-scale ceramics industries are able to produce a large volume at a low profit margin. For this reason, it is difficult for the small producer to compete directly. The small producer has an advantage of flexibility – he can produce a variety of products on demand and thus can supply local customers with special requirements.

For example, the modern tile industry is mostly very large-scale and can supply very cheap tiles of a uniform quality. The small producer can never compete directly with this. However, there is a growing market for specialty tiles, with decorations or relief designs, which the large producers cannot make. Many customers are interested in small quantities of special decorative tiles made according to their own design, even if the price per square foot is higher than mass-produced tiles.

Ceramics substituted for products made from other materials

In some countries, products like glasses for drinking tea may be produced more cheaply in ceramics. Or cement sewage pipes and toilet pans may be replaced by longer-lasting, more hygienic ceramic products.

4.3. STRENGTH REQUIREMENTS

Household items

Most common tableware items (cups, plates) can be made satisfactorily using either high or low temperature systems. Low temperature ceramics are more easily chipped and broken, but their low cost may be an advantage. High temperature products are stronger, and most hotels and restaurants will prefer them, unless the lower cost of earthenware makes up for the higher rate of breakage.

FIGURE 4.3.0.A Glazing electrical insulators. India.

Electrical insulators

Low tension insulators, fuse holders (kit-kats) etc. do not have to be very strong, so can be made in the low temperature range. High tension insulators have special requirements for porosity and strength, so must necessarily be made at high temperature.

Tiles

Glazed tiles are most commonly produced at low temperatures, which gives them sufficient strength for wall and floor applications.

Cold climates

Ceramic products to be used outdoors in freezing temperatures have special requirements, because of damage that can come from water freezing inside the product and causing it to break. These products are generally made at high temperatures, which make it possible to control water absorption.

4.4. INVESTMENT AND PRODUCTION COSTS

After considering the above decisions, the entrepreneur must then make an analysis of investment and production costs. These calculations are not easy to do, as the production of ceramics depends on so many complicated factors. For the new entrepreneur, it is important to start small and as simply as possible.

Low temperature systems usually require a lower initial investment, as kilns and burners will be cheaper. Fuel is usually the highest cost of production, and firing at low temperatures can save production costs. On the other hand, the cost of high temperature glazes is lower, as it is not necessary to use expensive frits.

In preparing a scheme for a new business, it is best to get help from a ceramics expert, who can help to figure out the comparative costs of the various options. Besides the usual overhead costs, it is necessary to consider:

- cost of clay body
- cost of glaze
- labor costs in production
- capital investment for equipment
- fuel costs
- working capital requirements.

5. SIMPLE GLAZE THEORY

5.1. BASIC CHEMISTRY

Chemistry is the science which describes what substances are made of and how they combine with each other. This science uses special names and symbols which are described below.

5.1.1. ELEMENTS/COMPOUNDS

Elements
An element is made of only one kind of *atom*. It cannot be broken down into more simple substances. Oxygen (O) is the most common element on earth.

Compounds
A compound is composed of more than one element combined chemically. Water (H_2O) is a compound made up of two atoms of hydrogen (H) and one atom of oxygen (O). Silica (SiO_2) is another compound and consists of one atom of silicon (Si) and two atoms of oxygen (O). This is the most abundant material in the earth's crust. Two or more atoms combined form a *molecule*.

Ceramic raw materials are usually in the form of *oxides*: an oxide is a compound that includes oxygen (O). Minerals are compounds.

5.1.2. SOLID, LIQUID, GAS

Solid, liquid and gas are the three states of matter. Most materials can exist in all of these states, depending on their temperature. A familiar example is water, which is *solid*

FIGURE 5.1.1.A Water is two elements combined. A molecule of water consists of two atoms of hydrogen and one of oxygen.

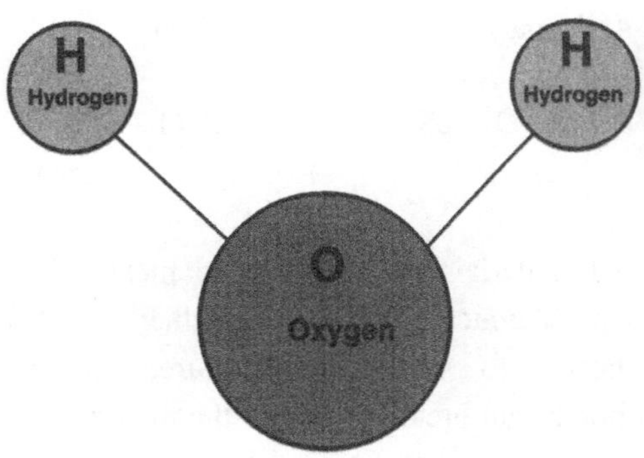

FIGURE 5.1.1.B A molecule of the compound silica (sand) has two atoms of oxygen and one of silicon.

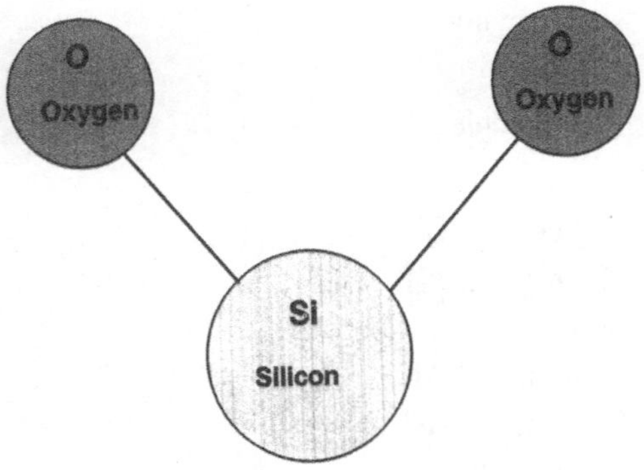

below 0°C, *liquid* from 0°C to 100°C, and *gas* above 100°C.

Making glaze depends on mixing *solids* together, applying them on a pot and then changing them to *liquid* in the kiln. Some of the glaze materials also become *gas* during firing and leave the glaze. On cooling, the glaze again becomes *solid.*

Mixture

A mixture is a physical, not chemical, combination of compounds (and sometimes elements) and each compound remains chemically unchanged in the mixture. Air is a mixture of oxygen, carbon dioxide, nitrogen and other gases. A glaze made of feldspar, quartz and lime is prepared by combining the compounds as a mixture, but during firing a chemical combination takes place and the fired glaze becomes a compound.

Chemical symbols

There are about 100 elements, and each of these has a name and a chemical symbol, which is used as an abbreviation of its name. Some of these symbols are the same as the first letters of the English name, but some are not!

For example:

Oxygen is "O"
Hydrogen is "H"
Silicon is "Si"
Alumina is "Al"
Sodium is "Na"
Lead is "Pb"

Compounds are written in a similar way with capital letters marking the individual elements: for example, water is "H_2O" and salt is "NaCl".

The small number "2" in "H_2O" indicates that there are two atoms of hydrogen for each atom of oxygen in water. If there is no number, it is understood that there is only one atom – so salt is one atom of sodium and one atom of chlorine.

The formulas of complex ceramic materials are written as compounds of oxides with a raised period (·) between them to show they are chemically combined. For example potash feldspar is written:

$$K_2O \cdot Al_2O_3 \cdot 6SiO_2$$

In the appendix the chemical formulas of other materials are listed.

5.1.3. CHEMICAL REACTIONS

The formation of clay from feldspar can be written in chemical symbols:

$$\underset{\text{(feldspar)}}{K_2O \cdot Al_2O_3 \cdot 6SiO_2} + \underset{\text{(water)}}{H_2O}$$

$$\rightarrow \underset{\text{(clay)}}{Al_2O_3 \cdot 2SiO_2 \cdot 2H_2O} + \underset{\text{(potash)}}{K_2O} + \underset{\text{(silica)}}{SiO_2}$$

All materials are built up of elements which are chemically bonded together. When heated to a high temperature, chemical bonds can break down and the material will change its properties. The production of quicklime by heating limestone to 900°C is an example of this:

$CaCO_3$	–> CaO	+ CO_2
(limestone)	(quicklime)	(carbon dioxide)

Carbon dioxide (CO_2) goes into the air, and the remaining quicklime (CaO) is slaked with water and can then be mixed with sand to form mortar for house construction. The mortar sets when the calcium oxide (CaO) takes back carbon dioxide (CO_2) from the air and thereby regains the hardness of the original limestone ($CaCO_3$):

CaO	+ CO_2	–> $CaCO_3$
(soft mortar)	(from air)	(set mortar)

5.1.4. SOLUTION/SUSPENSION

Solution
A solution is a mixture of molecules. For example, sugar completely dissolves in water: the separate particles consist of molecules of sugar and water. Sugar and water remain a solution until the water evaporates.

The higher the temperature of the liquid, the more solid material can dissolve in the liquid. When no more solid can be dissolved the solution is called "saturated".

Suspension
In a suspension the particles are bigger than molecules. A mixture of clay and water is a

FIGURE 5.1.4.A Glazing of a teapot by dipping it into a suspension, a mixture of glaze and water. Notice the practical turning table for newly glazed pots.

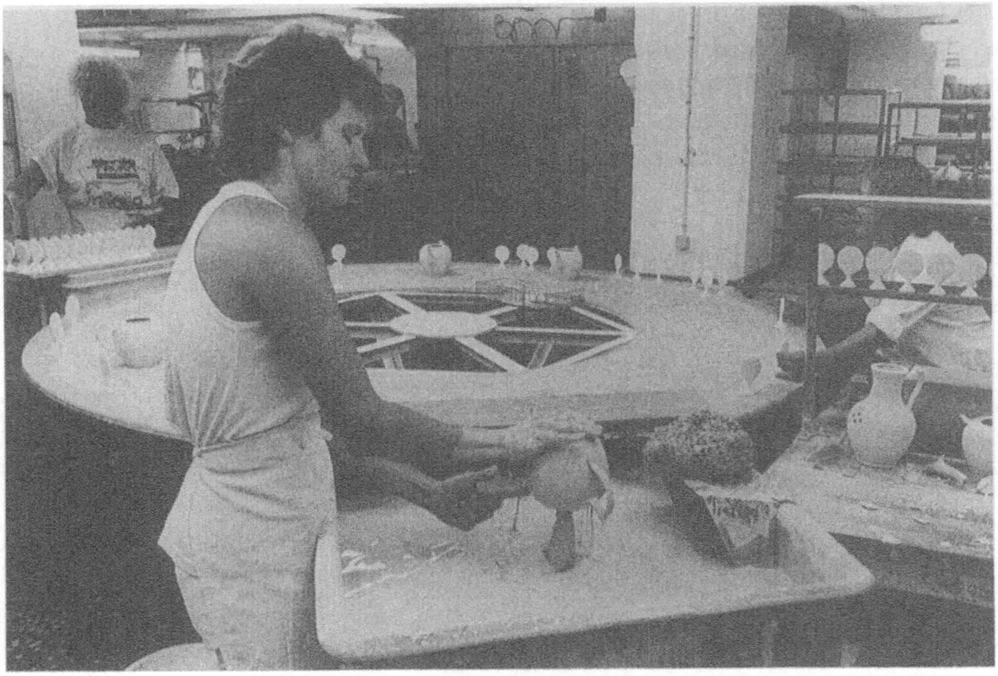

suspension. The clay particles are not changed by the water, and after some time the clay will settle at the bottom of the vessel. The clay is *insoluble* in water.

5.1.5. CRYSTAL STRUCTURES

Crystal structure

If we heat water to 90°C and add salt (NaCl), it will become dissolved in the water. If we continue to add salt until no more salt can be dissolved, the suspension is *saturated* with salt. If we let the solution cool to room temperature (20°C) the water can hold much less salt in solution, with the result that some of the salt will separate in the form of salt crystals.

All minerals have the form of crystals. When the water cools, the excess salt molecules start to combine with one another in regular patterns like small building blocks. The way the salt molecules connect to one another is very orderly and produces a cube-shaped crystal. Different materials will produce crystals of different shapes. The shape of a mineral's crystal is used to identify it.

FIGURE 5.1.5.A The cubic shape of a salt crystal.

5.2. GLAZE STRUCTURE

Glaze is similar to glass. Making glazes is confusing because there are so many raw materials that can be used. However, all of these raw materials can be broken down into three categories:

- flux
- glass former
- stabilizer.

All glazes require these three components. The main glass former is silica, the main stabilizer is kaolin, and the rest of the glaze is composed of one or more fluxes.

5.2.1. GLASS STRUCTURE

Silica (SiO_2) alone will make an excellent glaze if it is fired to its melting point (1715°C). Since this temperature is too high for ordinary kilns, other materials are added to lower the melting point of silica. Quartz is a crystalline form of silica found in nature. If a glaze forms quartz crystals when it cools, it will not be transparent, since light is refracted in many different directions by the crystal faces. Because glass or glaze is not usually crystalline, this does not happen.

A glaze or glass is a mixture of compounds that melts when heated. The melted liquid glass is like a solution. When the liquid cools, crystals start to form in a similar way as in a salt solution. However, the liquid glaze is very viscous (meaning sticky and semifluid) and the molecules cannot easily move around to form a regular crystalline pattern. So normally no crystals form during cooling, and the glaze remains clear like a liquid.

Glaze is, therefore, like a solid solution and is sometimes called a *supercooled liquid.*

5.2.2. FLUXES

Fluxes are the materials which lower the melting point of a glaze. They can be called *melters.*

Silica melts by itself but at a very high temperature. Therefore it needs additions of flux to make a practical glaze. The most common flux for temperatures below 1100°C is lead oxide (PbO), but since it is poisonous it is no longer used in modern crockery glazes. Another powerful flux is boron or boric oxide, B_2O_3, which is not poisonous and is used in glazes in the form of borax or boric acid. There are many other fluxes which contribute various properties of hardness, opacity, color response etc.

Fluxes are also called basic oxides or network modifiers.

5.2.3. GLASS FORMERS

Silica forms the main part of all glazes and is called a glassformer. The other glassformer is boron. Silica and boron are the building blocks of a glass or glaze. Other materials are only used to modify their behavior in the glaze.

Titanium oxide (TiO_2), tin oxide (SnO_2) and zirconium oxide (ZrO_2) also belong to this group. Sometimes they are called the acidic oxides or network formers, or the acid portion of the glaze.

5.2.4. STABILIZERS

Aluminum oxide, Al_2O_3, is added to make the melted glaze stiffer, so that it will not run off the pots during firing. It is called a *stabilizer.* Other words for stabilizer are: amphoteric, neutral or intermediate oxide.

Aluminum oxide has a high melting point and will increase the melting point of the glaze. It is usually added to the glaze as kaolin (china clay).

(Boron is termed a stabilizer in the USA but a glass former in Europe.)

5.3. EFFECT OF HEAT

As heat is increased, the molecules in the glaze move faster, resulting in drying, sintering, melting and gas escape. All of these effects occur when the glaze molecules move so fast that they start to break down, releasing some of their atoms and combining with other molecules to form the glaze.

5.3.1. DRYING

When the powdered glaze on the surface of the ceramic ware is heated, the water evaporates above 100°C (no matter how dry the glaze seems to be, there will always be some water remaining in it). The glaze layer should be as dry as possible before setting in the kiln. If the glaze layer dries too fast when firing starts, it may crack. This can cause crawling of the glaze after it melts.

5.3.2. SINTERING, MELTING, GAS ESCAPE

Sintering

As the temperature rises above 600°C, the sintering of the glaze powder starts. Sintering also takes place in the clay at this temperature. Sintering means that the glaze (or clay) particles start to stick to one another where they touch. The finer the glaze particles are ground, the earlier the sintering will start and the stronger the bond will become.

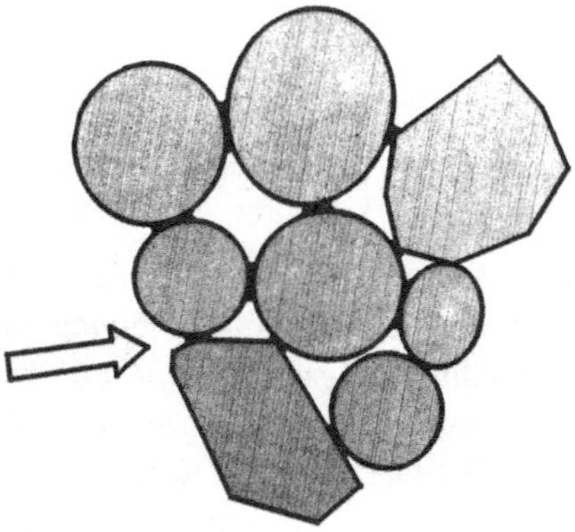

FIGURE 5.3.2.A The glaze particles are enlarged many thousand times showing sintering in a glaze heated to 600°C. At the points of contact (arrow) a weak bond is formed.

Fusion

As the temperature rises further, the most fusible (easy melting) materials in the glaze start to melt. This is called *fusion*. The refractory (hard melting) particles are surrounded by the liquid materials and are slowly included in the liquid.

The temperature at which melting starts depends on the materials in the glaze. Silica alone melts at 1715°C, but with additions of other materials the melting point will go down. Aluminum oxide (Al_2O_3) melts at 2050°C and calcium oxide (CaO) at 2570°C, but a mixture of 62% silica, 14.75% aluminum oxide and 23.25% lime melts at only 1170°C. A mixture which has a lower melting point than any of the single materials in the mixture is called an eutectic.

A mixture with many different materials will form eutectics (and will melt) at a lower temperature. Fine grinding of the glaze materials and prolonged firing time above the sintering temperature will also lower the melting point.

When fusion starts, the compounds also start to change. The chemically bonded water in clay has already been released. Around 900°C, limestone ($CaCO_3$) releases carbon dioxide (CO_2) and so do other materials containing carbonates, like barium carbonate ($BaCO_3$). Gases of sulfates, oxides etc. are also released both from the glaze and from the body. These gases have to pass through the glaze layer. This action mixes the glaze, helping it to become homogeneous.

In the beginning the melted glaze is very stiff (high viscosity), but as the temperature keeps rising the glaze becomes more fluid and, when watching the melting glaze surface through a spyhole in the kiln, bubbling or even boiling can be seen. When the glaze reaches its maturing temperature, the reactions stop and the glaze becomes smooth.

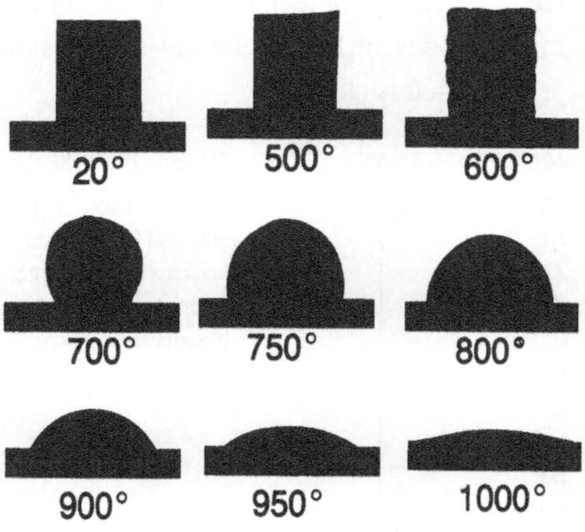

FIGURE 5.3.2.B A cube of glaze is gradually heated up to 1000°C. At 500°C the glaze shrinks slightly (sintering), but at 600°C it swells as gases develop. Melting starts before 700°C and is completed at 1000°C.

5.3.3. MATERIALS WHICH INCREASE/LOWER MELTING POINT

This chart shows the oxides according to their influence on melting temperature:

OXIDES WHICH RAISE MELTING TEMPERATURE

Al_2O_3	High ↑	
SiO_2		
MgO		
Cr_2O_3		Note this scale is not linear and depends on firing temperature and amount of oxide in the glaze
SnO_2		
ZrO_2		
NiO		
Fe_2O_3		
TiO_2		
CaO		
ZnO		
BaO		
FeO		
CoO		
CuO		
MnO		
PbO		
B_2O_3		
Na_2O		
K_2O	↓	
Li_2O	Low	

OXIDES WHICH LOWER MELTING TEMPERATURE

5.4. MELTED GLAZE BEHAVIOR

Fluid state

The fluid state of the glaze should be maintained long enough to allow all bubbles time to escape, so the glaze layer can heal over the holes left by the escaping bubbles. If a glaze tends to produce pinholes and craters, it can be given a soaking period (keeping the kiln at maturing temperature for some time) or the firing temperature can be raised in order to make the glaze more fluid (reduce viscosity).

If the glaze is too fluid, it will run off the pot or the fluid glaze will soak into a porous body leaving matt, dry spots on the surface.

The following chart shows materials which increase or decrease viscosity.

MATERIALS THAT INCREASE VISCOSITY

Al_2O_3	High ↑	Materials at top increase viscosity most.
ZrO_2		
SiO_2		
Cr_2O_3		Note these materials are mainly stabilizers and glass formers. (Scale is not linear.)
SnO_2		
NiO		
Fe_2O_3		
TiO_2		
CaO		Most materials in this group are fluxes.
MgO		
ZnO		
SrO		Materials at bottom decrease viscosity most.
BaO		
CoO		
MnO		
PbO		
K_2O		
Na_2O		
B_2O_3	↓	
Li_2O	Low	

MATERIALS THAT DECREASE VISCOSITY

5.4.1. SURFACE TENSION

To understand surface tension, fill a glass with water to the rim and look at the water surface. The middle of the water surface will be higher than the rim, but the water will not run over. The surface tension of the water holds it as if it were held by a plastic membrane.

A small amount of water forms a spherical drop. Larger amounts of water flatten the spherical form because the force of gravity increases with the weight of water. The fluid glaze behaves in a similar manner, and if the surface tension of the fluid glaze is too high the glaze will pull itself into small islands, leaving the clay body uncovered. This is called crawling.

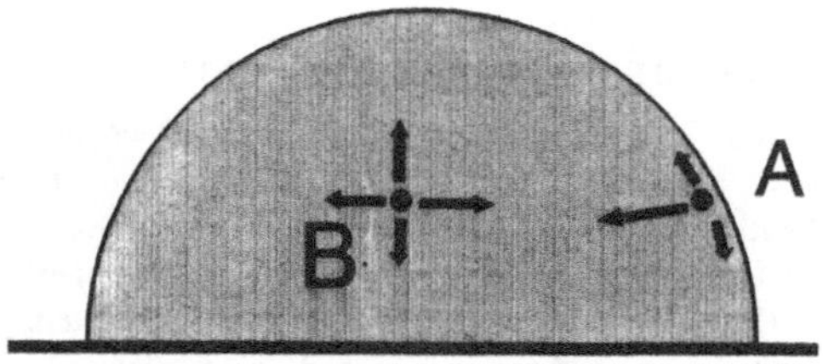

FIGURE 5.4.1.A Surface tension is created by the difference of forces acting on water in the center (B) and at the surface (A). A water particle at B has forces of attraction of the water around it evenly distributed. But at A the force is mainly directed away from the surface. This difference causes water to form itself into spherical drops.

Increasing temperature lowers the surface tension as Fig. 5.3.2.B illustrates. At 800°C the glaze forms a half globe but at 1000°C it has completely flattened out. Different ceramic oxides influence the surface tension as listed in this chart:

MATERIALS THAT INCREASE SURFACE TENSION

MgO	High ↑	Note the scale is not linear and the sequence of oxides may change due to other factors like viscosity, flue gas.
Al_2O_3		
ZrO_2		
ZnO		
CaO		
SnO_2		
Cr_2O_3		
NiO		
BaO		
SrO		
Fe_2O_3		
SiO_2		
TiO_2		
Li_2O		
Na_2O		
K_2O		
B_2O_3	↓	
PbO	Low	

MATERIALS THAT DECREASE SURFACE TENSION

5.4.2. CRAWLING

Crawling is caused by two factors:

- high surface tension of the glaze;
- difficulty for the glaze to stick to the body.

If the body surface is greasy or dusty the problem is aggravated. Crawling may also happen if the glaze layer cracks before it is sintered. This happens if the glaze contains a high amount of clay or has been ground for too long in the ball mill. The surface tension will then pull the glaze away from the cracks.

FIGURE 5.4.2.A Crawling.

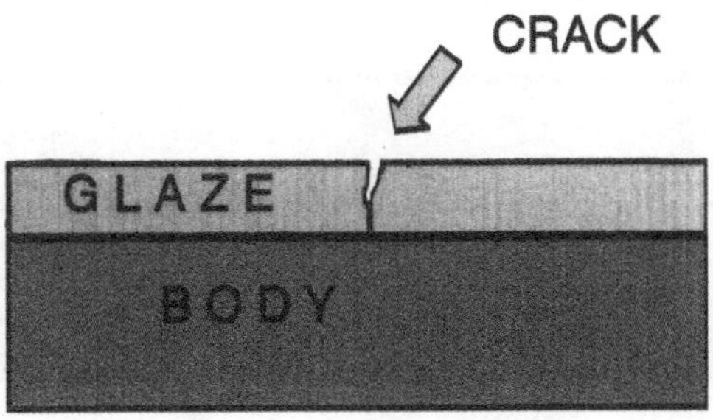

a) An enlarged section of body and glaze shows that a crack has developed in the dry glaze layer.

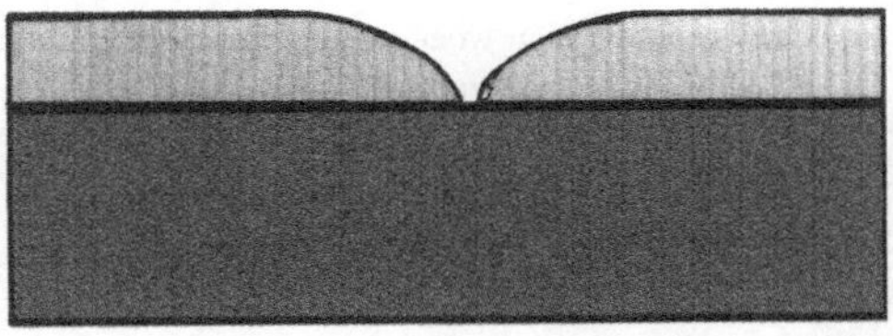

b) Glaze starts to melt and surface tension causes the glaze to pull away from the crack.

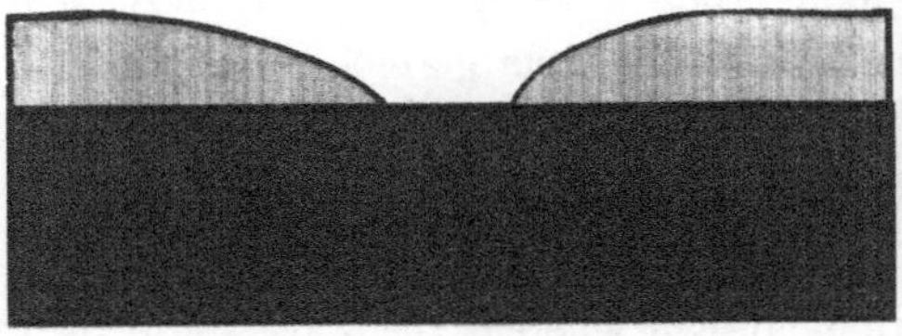

c) The body is left exposed. If the temperature is raised further the surface tension will be lower and the glaze might flow back and cover the body.

5.4.3. CRATERS, PINHOLES

The lower the surface tension, the shinier the surface of the glaze becomes and the easier it is for the glaze to heal over craters, bubbles and pinholes.

Interesting effects can be obtained by applying glazes with different surface tensions on top of each other (see page 80).

Surface tension, viscosity and melting temperature are interrelated, so when replacing materials all three will be affected.

5.5. INTERFACE BETWEEN GLAZE AND BODY

During firing the glaze interacts with the clay body. Some of the glaze will sink into the body and some of the body material will mix with the glaze so that an intermediate layer is formed between the body and the glaze. This layer bonds the clay and glaze together. It is called the glaze/body interface or "buffer" layer.

FIGURE 5.5.0.A Interphase layer created during firing by mixing of materials in the body and the glaze.

Effects of interface

Some of the coloring oxides in the body may enter the glaze and change its color. The higher the firing temperature the stronger the interface layer. The interface layer produces a strong bond between glaze and body that reduces the tendency to craze or peel.

Glazing on greenware (raw glazing or green glazing or single firing) promotes interaction between body and glaze. If too much of the glaze's flux combines with the refractory materials in the body, the glaze may become matt or dry.

5.6. COOLING AND CRYSTAL FORMATION

Glaze or glass is called a supercooled liquid because, during cooling, crystals have no time to form in the rather sticky mass, and glass by definition does not contain crystals. But some matt glazes and opaque glazes depend on the formation of crystals. For these, cooling should be slow to allow the crystals to grow. ZnO, BaO and TiO_2 are used for making matt glazes, but if cooling is rapid the glaze will become glossy instead of matt.

To avoid crystal formation, glossy transparent glazes should be cooled quickly after the maturing temperature has been reached.

5.7. TRANSPARENCY AND OPACITY

Transparency is the property of allowing light to pass through the glaze to the clay below. Transparent glazes may be colorless or have color in them – transparent blue, green, brown etc. It is necessary to use transparent glazes in combination with underglaze decoration. Transparent glazes are always shiny.

Opacity is the property of *not* allowing light to pass through the glaze. Colorless opaque glazes usually look white or gray. When coloring oxides are added, they can be any possible colors. They generally are used with overglaze or on-glaze.

It is possible to make glazes with every degree of transparency or opacity, such as semitransparent or semiopaque.

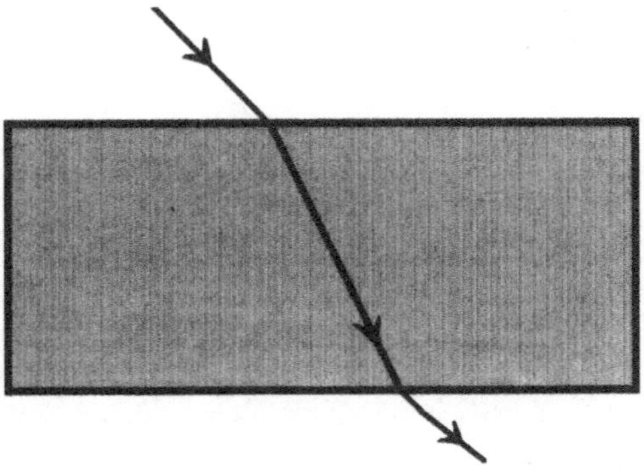

FIGURE 5.7.0.A Section of a window glass. A beam of light passes through it - it is transparent. The lights direction is slightly bent when passing from one medium (air) to another (glass). This is called refraction.

5.7.1. REFRACTION OF LIGHT

Transparency and opacity are determined by the glaze's ability to transmit light. When light strikes a transparent glaze, most of it passes through the glaze layer to the clay underneath, and the color we see is determined by the color of the clay. Thus, a transparent glaze on a brown clay body will look brown whereas the same glaze on a white clay body will look white. If the transparent glaze is colored, the clay body color will be changed by the fact that the glaze is green or blue, etc.

FIGURE 5.7.1.A A transparent glaze reflects the color of the underlying body.

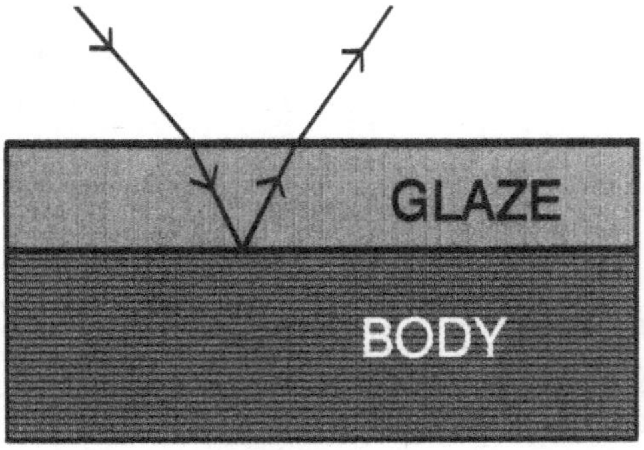

Opaque glazes have a large number of particles in them that reflect light, without allowing it to pass through the glaze. So we are not able to see through the glaze. Thus what we observe is only the surface of the glaze, which is not affected by the color of the clay underneath.

Semitransparent glazes have smaller numbers of light-reflecting particles, so they look cloudy or milky, and their color will be affected by the clay color underneath.

Transparent glazes can be made opaque by the addition of opacifiers, which are finely ground particles that do not enter into the melting of the glaze. These particles stay suspended in the glaze and reflect light. This is similar to mixing clay with water, which makes the water opaque.

Opaque glazes cannot be made transparent without changing their formula (unless they are transparent glazes with opacifier added).

The causes of opacity in glazes can be divided into 4 groups:

1. Presence of very fine particles, which do not dissolve in the glaze melt. The light going through the glaze is scattered by the fine particles. Tin oxide (SnO_2) and zircon ($ZrSiO_4$) are used for this.

FIGURE 5.7.1.B Fine particles of zircon or tin oxide in the glaze scatter the light and produce opacity.

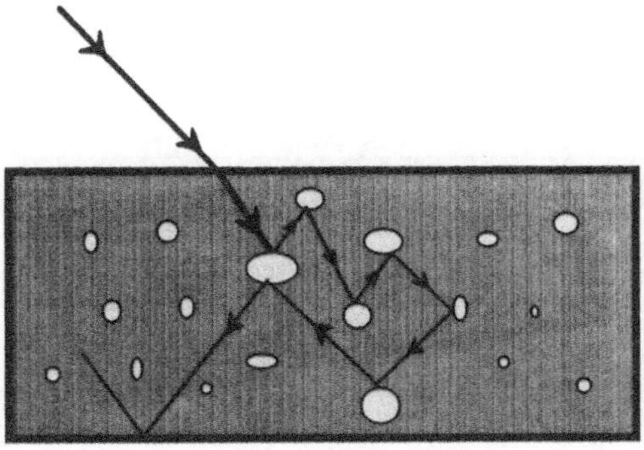

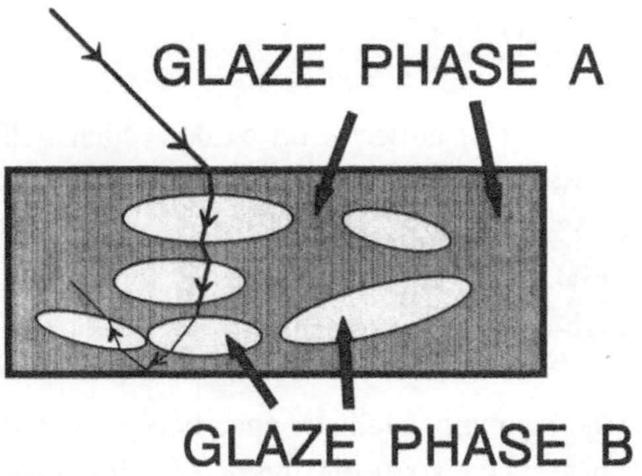

FIGURE 5.7.1.C Two glaze phases, A and B, in the melt cause opacity. Both glaze phases may be transparent but the light gets lost passing from one phase to the other.

2. Crystals formed in the glaze during cooling will scatter the light, causing opacity. Titanium dioxide (TiO_2) recrystallizes if the cooling is slow and can make glazes opaque.

3. Opacity is also caused when two melting phases of the glaze do not mix. The light will be scattered when it passes through the border between the two different melts. This takes place in boron glazes and with calcium phosphate (bone ash).

4. Gas bubbles scatter the light and produce opacity. This type of opacity is difficult to control and the method is not recommended.

In practice, a combination of the four methods is used. For example, an opaque glaze can be made with boron and additions of lime, zinc oxide and zircon.

5.7.2. MATERIALS CAUSING OPACITY

The best opacifier is tin oxide, which will make most glazes opaque in additions of up to 7%. However, it is a very expensive material and today is only used for special high-cost products.

Commercially available opacifiers are based on zirconium silicate, prepared with other additions such as magnesia and zinc oxide. They are marketed under names such as "zirconium opacifier", "zirconium silicate", "zinc zirconium silicate" and "magnesium zirconium silicate". Most of these are added to glazes from 5 to 10% and produce different results depending on the type of base glaze. They also vary widely in quality, and it is important to test them before ordering a large quantity. Zirconium opacifiers have the disadvantage of making glazes more refractory and often cause pinholing problems.

The main opacifiers are: (see also page 114)

Tin oxide, SnO_2
Zircon, zirconium silicate, $ZrSiO_4$
Titanium dioxide, TiO_2
Alumina, Al_2O_3 (high content in boron glazes will reduce opacity)
Calcium oxide, CaO (improves opacity in boron glazes)
Zinc oxide, ZnO
Calcium phosphate, bone ash, $Ca_3(PO_4)_2$.

Particle size
The finer the particle size of the opacifier, the better it works. Zircon is often included in the frit batch for greater opacity. In this way opacity is obtained with less zircon, thus reducing some of zircon's bad side effects like high viscosity and the tendency to cause pinholes. Unfortunately the addition of zircon to the frit increases its melting point, making it more difficult to run it off the frit kiln. It also increases the hardness of the frit so much that it may be difficult to grind it with ordinary pebbles and ball mill lining.

It is important to make sure that the opacifier is well dispersed in the glaze. The fine particles tend to lump together. This reduces the opacity effect. By ball milling the opacifier together with the glaze a good dispersion is assured.

5.8. SHINY OR MATT GLAZE

Glazes are also defined by the way they reflect light: they may be shiny or matt or in between.

Shiny
Shiny glazes are also known as "glossy" or "bright". They have the property of reflecting light like a mirror. They are best for utilitarian wares, sanitary ware and insula-

FIGURE 5.8.0.A A glossy glaze with a smooth surface reflects the light without scattering it.

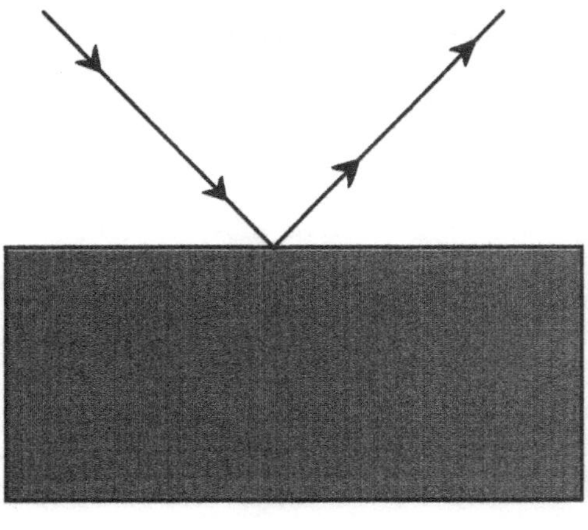

tors, as they are easy to wash and do not scratch easily.

Matt

Matt glazes are also known as "dull" or "nonreflective". Their surface can vary from smooth to very rough. They are useful for decorative wares and are very popular for floor tiles, which need to be beautiful but not slippery. The matt surface is not functional for dinnerware, because used with cutlery it makes an unpleasant sound and scratches easily.

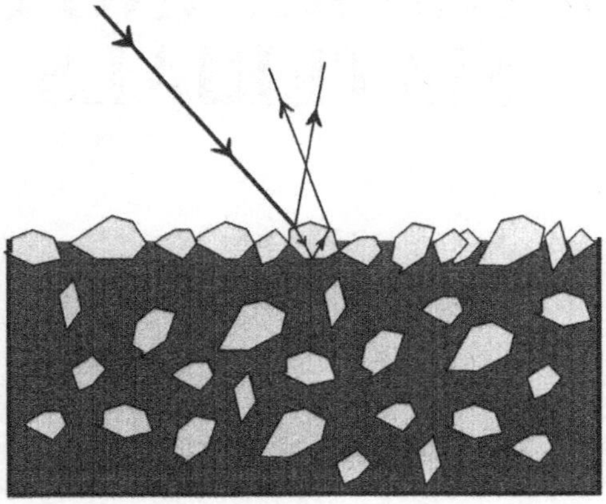

FIGURE 5.8.1.A Surface of a crystal matt glaze enlarged several hundred times. Crystals in the glaze scatter the light by sending it in many different directions.

5.8.1. MATERIALS CAUSING MATTNESS

There are several ways to produce a matt glaze:

Underfiring

As glaze begins to melt, it becomes glassy. If the firing is stopped before the glaze is completely melted, even glossy glazes will appear matt. Often these underfired glazes will have other problems such as blisters and pinholes, but some glossy glazes make very good matt glazes if fired a few cones below their normal temperature. Similarly, adding refractory oxides to a glaze (such as china clay or calcium carbonate) will produce a matt glaze that really is just an underfired glossy glaze.

Crystalline matt

Crystalline matt glazes develop small crystals which break up light (Fig. 5.8.1.A). This type of matt glaze usually produces a more smooth surface than underfired matt glazes. Some matt glazes depend on slow cooling to have time for the crystals to develop.

Barium carbonate, zinc oxide, titanium dioxide, magnesium oxide and calcium oxide are the agents for crystal matt glazes. For more details see page 113.

5.8.2. OTHER CAUSES

Sometimes glazes that should be glossy will become matt. Some of the reasons are:

- Some of the flux materials may evaporate during firing, leaving a matt surface.
- Sulfates from fuel may settle on the surface of the glaze.
- The glaze is applied too thin.
- The glaze was not mixed sufficiently or not sieved finely enough.

6. OBTAINING GLAZE MATERIALS

6.1. MATERIALS SUPPLIERS

In countries with large ceramics industries, there are suppliers that specialize in collecting and distributing raw materials. These may be mining companies that can supply specific items like clay and feldspar. If these can be obtained directly, it saves the costs of middlemen. However, these companies often deal only in large quantities. For the small producer, it is often best to get supplies from reliable distributors.

6.1.1. LOCAL SUPPLIERS OF CHEMICALS

General chemical suppliers or pharmacies often have many of the necessary ingredients for glazes (which are often used in other industries). They are useful for obtaining small amounts of chemicals, but often their prices are high.

6.1.2. SUPPLIERS OF OTHER INDUSTRIES

Glaze materials are often available from other types of suppliers. For example, agricultural suppliers can provide calcined limestone. Paint industries use materials such as iron oxide and opacifiers.

6.1.3. IMPORTED MATERIALS

Imported materials should only be considered if there are no local sources, as they are expensive and often customs and import regulations make it difficult or impossible for the small producer to obtain them. On the other hand, it is often worth paying the additional price, if it makes possible production of special glazes or decoration effects that are in demand in the market.

In Thailand, for example, where there is a large export market for decorative ceramics, many producers import clay, glazes and overglazes from Japan. Their profit comes from cheap labor and high value added.

6.2. MATERIALS FROM NATURAL SOURCES

Small producers can mine their own materials if these are available in the area. Historically, pottery centers located themselves where the necessary clay and glaze materials were available. Where stoneware clay and high temperatures are used, it is possible to make glazes from low-temperature clay alone. Generally, stoneware glazes are made from the basic ingredients of feldspar, quartz, limestone and clay, which are quite common. Wood ash is an-

FIGURE 6.2.0.A Traditional terra cotta is decorated by rubbing gas welding wire on the surface of the green pot. After firing the decorations turn dark red.

other common base for high temperature glazes. The process of mining, selecting and grinding is quite time-consuming, and with the advent of modern transportation it is often cheaper to purchase materials from suppliers.

In Nepal, we developed low-temperature glazes based on borax, which must be imported. The bulk of the glaze is composed of local materials such as rice husk ash (for silica), limestone and local clay, which are all easy to get and cheap.

6.2.1. CRYSTAL ROCKS

Igneous rocks

When the young earth slowly started to cool, different minerals formed crystals in the mass of molten rocks (magma). A variety of crystalline rocks were formed differing in composition according to their locality. For example, the igneous rock called basalt was created at a great depth and contains little feldspar compared to granite, which formed near the surface.

FIGURE 6.2.1.A Crystals of quartz. The large crystal is 15 cm long.

If rock cools very slowly, crystals have time to grow large, whereas rapid cooling produces small crystals. This process is still going on today where movement in the crust of the earth causes deep layers of molten materials to rise to the surface. An erupting volcano lets out hot magma, which cools quickly. The resulting volcanic rocks have microscopic-size crystals, since the rapid cooling allows little time for crystals to grow.

The most common crystal rocks used in glazes are feldspar and quartz. If a piece of granite is picked up and broken in two, the fresh faces of the stone will show a shiny surface and the crystals of the different minerals can be identified. The black crystals are mica or tourmaline. The yellow, white or red colored crystals with a pearly shine are different types of feldspar. The clear colorless crystals are quartz. The weathered surface of the granite will most probably show a rough surface with many holes, where the soluble feldspar crystals have been washed away by rain, whereas the less soluble crystals of mica and quartz remain. Coarse granite (known as pegmatite) often breaks up in weathering, leaving large pieces of quartz and feldspar lying on the ground. These can be collected, ground and used in glazes.

Volcanic rocks

These are rocks formed by the action of volcanoes, often in the form of molten lava that flows out of the volcano. The crystals in the rock are extremely small because the lava cooled very fast. Lava is essentially a glaze and can be used as the basis of high temperature glazes.

6.2.2. SEDIMENTARY ROCKS

Sedimentary rocks are made of materials produced by the crumbling of old rocks. All rocks eventually break up in the course of time when exposed to weather, and the broken-up rock particles are carried away by water. These particles of clay and sand are transported to lower lying areas or to the

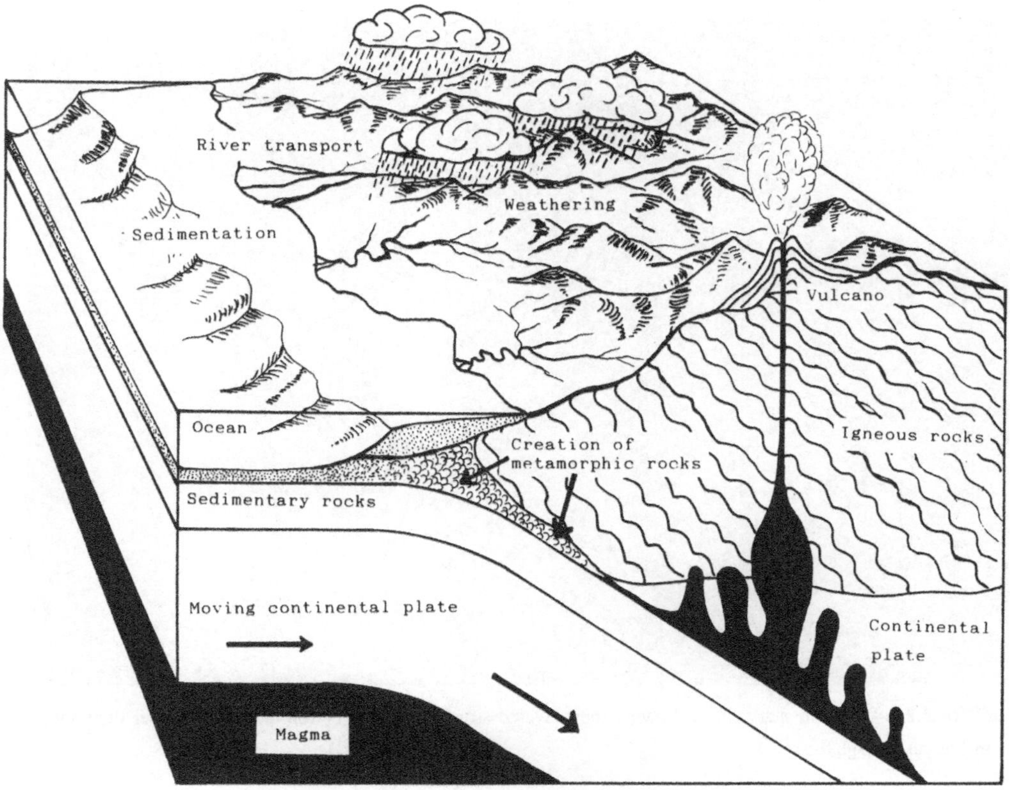

FIGURE 6.2.1.B A cutout of a section of the crust of the earth shows a continental plate moving under another. The friction of the plates generates heat, which melts rocks and feeds a volcano. Rain falls and old rocks are weathered and washed to the sea creating new layers of sediments. Later the sediments are compressed into rocks.

sea where they settle one layer upon the other. In the span of millions of years, the growing weight of sediments causes the deeper layers to compact and gradually turn into rocks, called sedimentary rocks. Much later, the movement of landmasses sometimes turns the whole area upside down, so that the old sea floor, with its sedimentary rocks, becomes a new range of mountains.

The upper part of new mountains consists of sedimentary rocks resting on deeply set igneous rocks. Sedimentary rocks like sandstone, shale and slate can often be recognized by their layered structure. Limestone is a sedimentary rock created by the skeletons of billions of small animals that lived in the ancient seas. Gypsum is formed by chemical sedimentation in areas where seawater evaporates on a large scale. This produces a high concentration of gypsum which forms crystals like the formation of salt crystals in a glass of salty water.

For the glazemaker, sedimentary shale can be a source of glaze. At high temperatures, shale melts and with a few additions will produce glazes that are usually brown. Al-

FIGURE 6.2.2.A Sedimentary rocks showing their layered structure. At the bottom a river that over thousands of years has cut through the rocks.

though shale often does not slake in water, it can be ground in a pan mill and used in glaze.

6.2.3. METAMORPHIC ROCKS

Igneous and sedimentary rocks are sometimes changed into new forms by high temperature and pressure. Marble is an example of a metamorphic rock formed from the sedimentary rock limestone.

6.2.4. HOW TO GET INFORMATION

Local authorities

First of all, information about the geology and the minerals of the region should be gathered from local authorities, like industrial development organizations, agricultural institutions, National Geological Institutes or mining corporations. They may have little information, and the authorities may even say that no materials are available in the region. However, that is often not true and should not keep anybody from looking on his own.

Practical people

It is worth talking to people who make water wells, and builders of dams and roads. They sometimes have useful information about the minerals of the region. Farmers in the area will know about the upper layers of soil on their fields and about local rocks. Sometimes glaze minerals are used for other purposes, like whitewashing houses or medicine.

The best source of information is often other potters.

6.2.5. LOOKING FOR MINERALS

Good places to look for minerals are in riverbeds, where many different types of rocks will wash down from the mountains above. Although most of these may not be useful, it is often possible to find quartz and feldspar. Any rock with an unusual color is worth testing. Rocks that are unusually heavy may contain metallic oxides. For the potter, however, there are few rocks that are directly useful, other than quartz, feldspar and limestone, and some of the volcanic rocks.

Other minerals that are useful in glazes are sodium and potassium compounds, which sometimes form on the edge of lakes, particularly in desert areas. These usually look like a white powder and are soluble in water.

6.2.6. TESTING

To begin with, the most useful test is to take a small sample of the material, place it in a clay bowl and fire it in a regular glaze firing. This will indicate if it melts or not. If it melts, it certainly can be used in a glaze. Materials that do not melt should not be automatically rejected, as many useful glaze materials (such as calcium carbonate and quartz) only melt when combined with other materials. The simplest way to find out if they are of use is to make a line blend of one of your standard glazes, combined with the unknown material.

Rock minerals can be identified by their crystal shape, color, specific gravity and hardness. If you are seriously looking for rock minerals there are good books presenting most common minerals with color photos.

Hardness

Mohs' scale of hardness is based on the hardness of 10 different minerals:

1	Talc
2	Gypsum
3	Calcite
4	Fluorspar
5	Apatite
6	Orthoclase feldspar
7	Quartz
8	Topaz
9	Corundum (pure alumina)
10	Diamond

Window glass and a penknife are about 5.5 and a metal file about 6.5.

Two materials have the same hardness if they cannot scratch each other. Quartz can scratch feldspar but not topaz. In the field a piece of glass and a penknife are used to find out if the hardness of a rock is higher or lower than 5.5.

Chemical analysis

If a testing laboratory is available, samples can be sent there for chemical analysis. This is usually expensive but may be helpful if the material looks useful after firing.

6.3. OTHER SOURCES OF MATERIALS

Recycled materials are often useful in glazes. These may be by-products from other industries, such as rice husk ash or bone meal, or waste materials. Some other sources of useful materials are discussed below.

6.3.1. METALLIC OXIDES

Metallic oxides are used as coloring agents in glazes. Commonly available are:

Iron oxide, which can be obtained by scraping rust from old steel. It is often possible to get this from paint and hardware suppliers, who use "red oxide" for coloring paint and cement.

Manganese dioxide, which is the main ingredient in torch batteries (the black substance which can be removed from old batteries).

Copper oxide, which can be collected from makers of copper pots. The oxide is the black powder that forms on the surface of copper when it is heated. Another way is to fire copper wire in the kiln and to use the resulting black copper oxide.

6.3.2. ASHES

Wood ashes are used as the basis for high temperature glazes, since they contain sodium, potassium, silica and other ingredients. Early glazes were often simple mixtures of wood ash and clay. Most wood ash is suitable for this purpose, but each type of wood will produce different characteristics and will have a different melting point. So it is important to have a consistent supply. Ash must be sieved to remove unburned material and is usually washed in water and dried before use. If it is not washed it contains more fluxes but they are soluble and make the glaze slip caustic.

At cone 8 to 11, a good starting point is 2 parts ash, 2 parts feldspar and 1 part clay. Ash glazes have the following general limits:[1]

Ash	20-70%
Feldspar	20-70%
Whiting	5-20%
Flint	15-25%
Clay	5-20%

Rice husk ash contains more than 90% silica, so it can be used instead of quartz in many cases. For accuracy, it should be burned white – if there is much black carbon in it, it will make calculations incorrect. In the Appendix the chemical composition of different ashes is given.

6.4. STORING, PACKAGING AND LABELING

If you use local materials, they will change from time to time. For this reason, it is best to store as much material as possible and to check each new batch by trying it in a standard glaze. For example, feldspar tends to be variable and, as the mine is used, the chemical composition will change. Suppliers of

[1] D. Rhodes, *Clay and Glazes for the Potter*

feldspar usually keep several large storage areas of material from different parts of the mine. In order to keep it uniform, they mix the different feldspars together when supplying.

Some materials are damaged by water. Borax, boric acid, soda ash and plaster of paris should all be kept in a dry place. In particular, soda ash absorbs water (up to 7% after one year, 11% after two years) and will thereafter no longer be effective as a slip deflocculant; and plaster will not set correctly after damp storage.

When you get local materials, each batch should be kept separately and labeled with date and source. It is often a good idea to purchase more material when your old supply is about 50% finished and to test it to see if it is the same or not. If it is not greatly different, the new material can be mixed with the old and your glaze will not change unexpectedly.

A good labeling system is very important, as most glaze chemicals look rather alike. Never depend on your memory – keep a permanent label on the bag or jar of material. Additionally, if you order bags of material from a supplier, ask him to label the outside of the bag, and also to put a label inside the bag as labels are often lost in shipping.

7. FRITS AND FRIT MAKING

Most low-temperature glazes require fluxes that are either poisonous (lead) or water-soluble (sodium and potassium). Traditionally, these materials were used raw, but this is not satisfactory for modern potters. Raw lead is poisonous and sodium/potassium are water-soluble. Borax is sometimes used raw in glazes, but these glazes cannot be stored for a long time, as the borax will go into solution or form crystals.

FIGURE 7.0.0.A Melted frit glass running out of a frit kiln into water.

The principle of fritmaking is very simple: molecules of poisonous or soluble fluxes should be chemically combined with glass-making materials to eliminate these undesirable characteristics.

A frit is a combination of a flux or several fluxes (lead, borax, boric acid, potassium carbonate) that is combined with other insoluble materials (quartz, feldspar, lime etc.), melted in a kiln to form an insoluble glass, and ground to be used as the base for making glazes. (Many low temperature glazes are simply 90% frit and 10% china clay).

Fritmaking is not usually practical for the small producer, as it takes time and requires a special kiln and a ball mill for grinding. On the other hand, if reliable frits are not commercially available, the potter may have to produce his own. Frit glazes are more expensive than raw glazes, but their convenience usually makes up for the additional cost.

There are many different commercially available frits, all designed for different temperatures, surface qualities, coefficients of expansion, and color responses. The potter trying to decide which frit to use must depend on the supplier, as formulas are usually kept secret. Suppliers will give advice on which frit is best for the potter's purpose. There are two main types of frit:

Lead frits

These are all designed to provide lead in a nontoxic form. Lead oxide is combined with other materials to give desired properties of surface, opacity, and color response. The standard lead frit is called lead bisilicate and is simply a combination of lead oxide and silica, which combines the lead in an insoluble form. This can be used as the base for a large variety of lead glazes.

Other frits used commonly in the tableware industry are called lead-borosilicate frits, which combine the desirable properties of both lead and boron and are generally safer to use.

WARNING! Lead frits can still be poisonous, and glazes made from them can be poisonous if they are not combined with sufficient silica to combine with all the lead molecules.

Leadless frits

These are based on boron compounds, again combined with other materials. Because glazes compounded with lead are difficult to control for lead release, leadless frits are recommended for small producers.

7.1. WHY MAKE FRITS?

Frit making is only suggested if reliable commercial sources are not available. Often frit manufacturers are not interested in supplying small amounts. Dishonest frit manufacturers sometimes sell bad batches of frit to small producers. Even though frit making is complicated, the small producer who makes his own frits at least has the process under his own control.

Raw borax glazes can be used, but they must be used immediately after mixing or problems will result from the soluble borax. This may be satisfactory for art pottery but, if consistent results are needed, it is better to use fritted glazes.

Similarly, raw lead glazes are widely used. This is a danger for the workers, who will eventually develop lead poisoning unless they take extreme care in handling the glaze. Modern industries never use raw lead glazes, and industrialized countries all have severe restrictions on the use of lead in glazes. In developing countries, workers in industry suffer from lead poisoning, and it is the responsibility of the industrialist alone to take care of the workers' health. As lead poisoning takes several years to develop, many factory owners do not understand the seriousness of the problem and continue to harm their workers. IT CAN TAKE UP TO 20 YEARS TO DEVELOP SYMPTOMS OF LEAD POISONING!

7.2. FRIT PRODUCTION

Frit composition

All the soluble materials are included in the frit batch along with silica, in order to form a glass when fired in the frit kiln. Other materials may be included for modifying the frit or helping to melt it.

The main frit raw materials are:

Silica sand, SiO_2
Rice husk ash, almost 95% SiO_2
Borax, or sodium borate, $Na_2B_4O_7.10H_2O$
Boric acid, H_3BO_3
Limestone, $CaCO_3$

Feldspar, soda and/or potash, K_2O & $Na_2O\cdot Al_2O_3.6SiO_2$
Clay, $Al_2O_3.2SiO_2$
Zinc oxide, ZnO
Zircon, $ZrSiO_4$ (opacifier)
Red lead oxide, Pb_3O_4
Other materials like talc, barium carbonate and bone ash may be added.

In order to have a frit with low viscosity that easily runs out of the kiln, the clay or alumina of the glaze is not added to the frit. However, in order to make the ingredients insoluble, 2-3% kaolin should be included in the frit. For more details on frit calculation see page 143.

Work flow

The work flow for frit production is shown in Fig. 7.2.0.A. It is better economy to prepare large frit batches when firing a continuous-type frit kiln.

Prepare materials

All materials for frit need to be clean, dry and ground to pass through a 60-100-mesh sieve. The finer the material, the easier it will be to melt it. If rice husk ash is used as the source of silica, it should be well-burned to a white color, so that unnecessary carbon is not introduced. If there is a large amount of black carbon, this will decrease the amount of silica available. The content

FIGURE 7.2.0.A Work flow of frit production

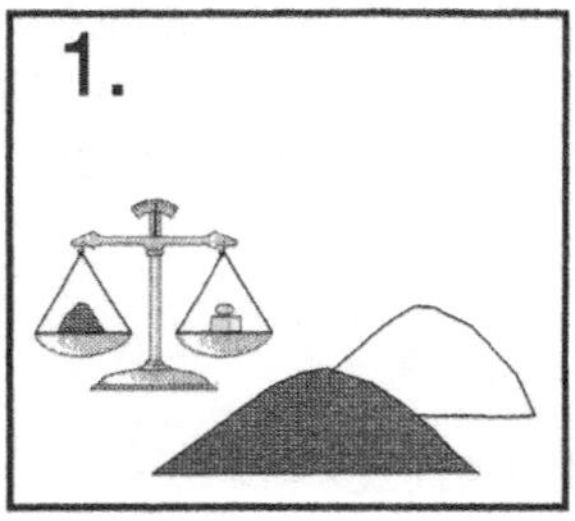

Weighing frit materials according to recipe.

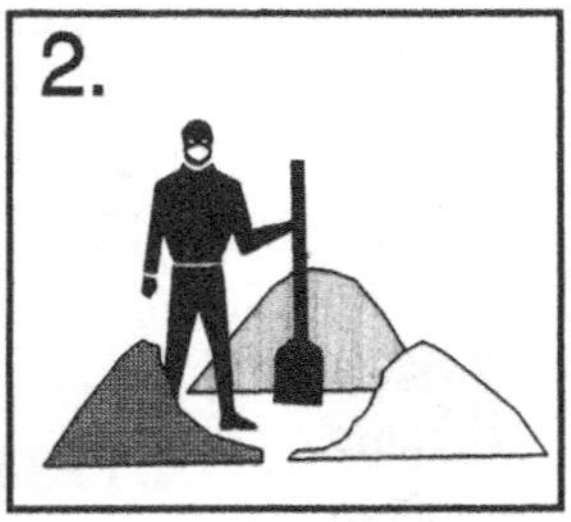

Dry-mixing the materials. Wear a dust mask!

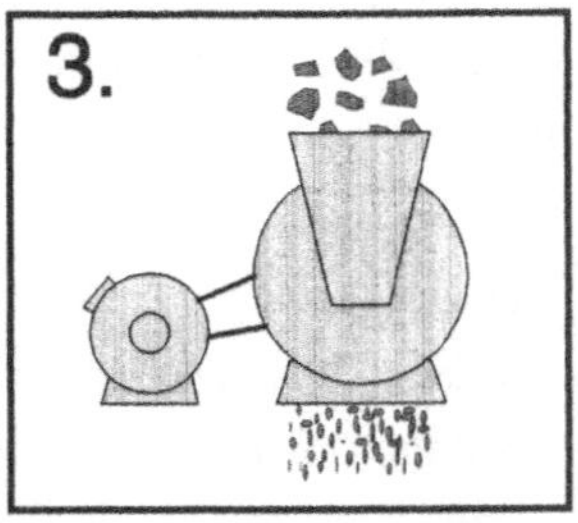

For thorough blending the materials are screened or mixed in a hammer mill.

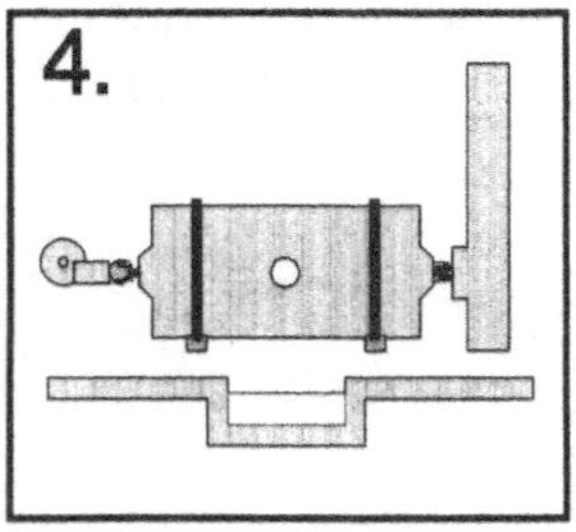

Melting in a frit kiln (here a rotary kiln).

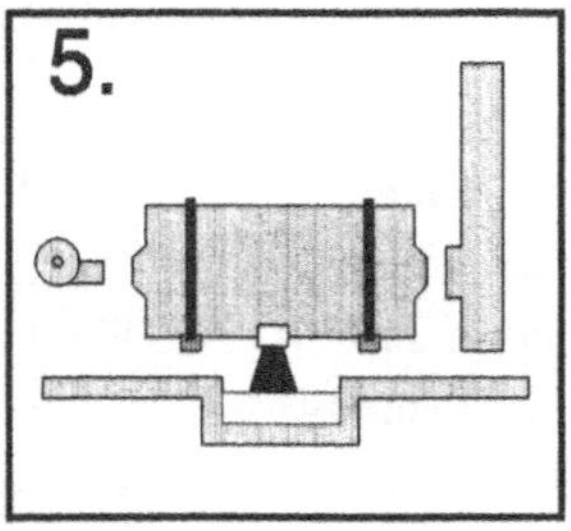

The melted frit is poured into water to form frit granules.

The frit is dried and filled in bags for later mixing with other glaze materials and ball milling.

of carbon in rice husk ash may vary more than 30% from batch to batch. If materials are wet, they should be dried completely so that the weight of water is not included in the recipe. In frit calculations, the loss on ignition (see page 146) needs to be included to account for loss of material during firing.

Blend materials

Weigh the materials accurately and blend them together dry. WEAR A DUST MASK! Small amounts can be mixed by hand in a bucket, and larger amounts can be mixed with a shovel on a clean cement floor. After mixing the frit materials they are screened through a 16-mesh sieve (mosquito net) to ensure thorough blending or the materials are run through a hammer mill.

Melt the frit in a kiln

There are many different systems for melting frit, which are described below in section 7.2. In each system, the principle is to thoroughly melt the frit until all ingredients are combined. Most frit is melted at 1150°C to 1250°C.

Check the frit

A sample of molten frit should be taken and examined to see if the melt is complete. The frit should be uniform, without particles of unmelted material.

With continuous frit kilns, the rate of feeding raw frit and the speed of the melted frit must be adjusted so that all the material melts completely and has time to mix properly.

FIGURE 7.2.0.B Frit glass drawn from kiln. The small granules to the right were produced by quenching in cold water. The lumps to the left were not cooled in cold water.

Quench the frit in cold water

The molten frit is poured into cold water, which "shatters" it into small pieces that can easily be ground. With continuous melting and discharging it is necessary to let fresh cold water run continuously.

Grind the frit

If the frit is quenched correctly, it will be easy to put it directly into a ball mill and grind it until it can be passed through a 100-mesh sieve. The granulated frit may be first dried and then stored in bags until it is needed for glaze making. Then it is ball-milled together with clay and other glaze materials. Alternatively, the still wet frit is ball-milled first.

Sieve the wet frit

When the frit is removed from the ball mill, it should be sieved through 100 mesh to remove any large particles that were not ground.

Dry the frit

The wet frit is settled, excess water is poured off, and the remaining frit can be spread out to dry, either in the sun or in a dryer.

Test the frit

Each batch of frit should be tested for correctness. The simplest way is to fire it in a kiln on a specially made flow tester, along with a sample of correct frit (page 131). If the frit flows evenly to the control sample, it will probably be correct but should be double-checked by trying it in a standard glaze.

Additionally, the frit should be tested for solubility in water. A sample amount is boiled in water for several hours, then allowed to sit for 2 weeks. If crystals do not form during this time, the frit can be considered stable. If crystals form, it means that there is not enough silica/alumina in the frit and the composition will need to be changed (frit calculations page 144). The causes of crystal formation could also be with the frit firing, e.g. overcharging, too short a firing time and improper mixing.

The finished tested frit may be sold to other ceramics producers either as a milled powder or in granular form.

7.3. FRIT KILNS

There are many different kinds of frit kilns, which are selected according to the amount of frit that needs to be regularly produced.

Normally, each type of frit – transparent, opaque, lead – requires a separate kiln to prevent contamination. When one kiln is used for several frits, it must be cleaned out before each different batch by melting frit in it to remove most of the old batch. This contaminated frit is then kept separately, to be used as "clean-out" frit before changing to different compositions.

7.3.1. CRUCIBLE FRITTING

Small amounts of frit for testing are easily made in a fireclay crucible. The crucible with frit is fired together in a glaze firing, which will melt the frit into a solid block of glass. After firing, the crucible is broken away from the frit and the frit can be crushed and ground. It is a good idea to first paint the inside of the crucible with china clay slip, as this will make it easier to separate the frit. NOTE: Frits containing boric acid often cannot be melted successfully this way, as the boric acid melts at a very low temperature and flows to the bottom before the rest of the ingredients melt. Frits with rice husk ash may also be difficult to melt in this way, because the upper layer of the frit melts first sealing off the frit mixture so that the carbon remaining in the ash cannot burn out. Carbon is highly refractory and it will prevent the frit from melting.

This is only suitable for test production and is not a safe method, since the pot often cracks, resulting in frit running out, destroying other ware, kiln furniture and the kiln lining.

CAUTION: Borax frits boil during melting with a great increase in volume. The crucible should be filled only half with frit, and a tile placed over the top to prevent boiling over.

7.3.2. CRUCIBLE KILN

For fritting small amounts of frit a simple frit kiln is shown in Fig. 7.3.2.A. It can be fitted with several crucibles arranged in a row for melting different frits at the same time. The crucibles can be loaded with raw frit from the top. The fuel economy of this type of kiln is less than for the other kilns.

7.3.3. OPEN HEARTH KILNS

Open hearth kilns consist of a tank made of firebricks, which is set in a crossdraft kiln. The kiln may be fired by coal, firewood, oil

FIGURE 7.3.2.A Coal-fired frit kiln with three crucibles.

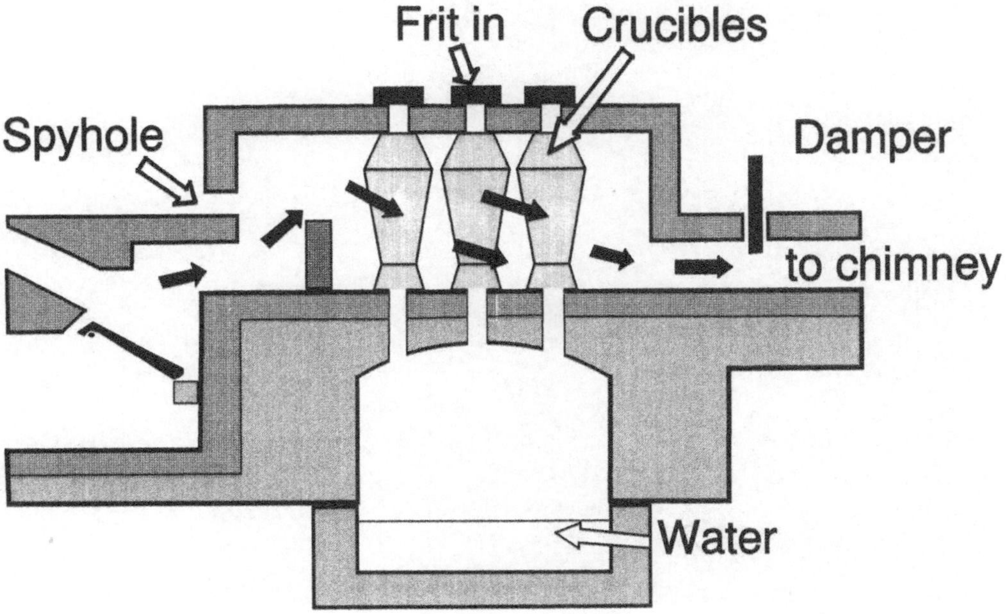

or gas. The hot flue gases heat the arch over the frit. The arch in turn heats the frit. In batch-type frit kilns, the frit melt is checked by drawing out some melted frit with an iron rod for inspection.

After the frit is completely melted, a hole at the bottom of the tank is opened and the frit flows out into cold water. Then another batch of frit may be charged from an opening in the arch.

The melting of several tonnes of frit may take 6-12 hours consuming 1-1.5 tonne coal per 1 tonne melted frit.

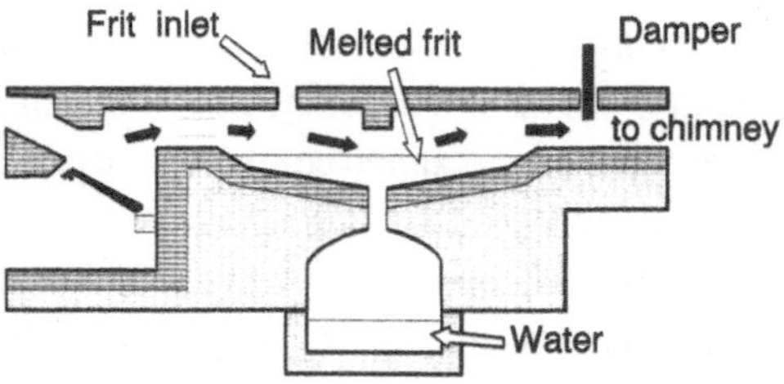

FIGURE 7.3.3.A Open-hearth frit kiln for coal firing.

FIGURE 7.3.4.A Continuous-flow frit kiln with one wall removed. This design used a cast cover and raised kiln slabs for the slope. A firebrick arch proved more durable and the raised slabs did not increase fuel efficiency only the cost of construction.

7.3.4. CONTINUOUS FLOW

The continuous-flow frit kiln uses a kiln with a sloping floor, made of fireclay refractories. The raw frit is introduced at the upper end and, as it melts, it flows down while mixing to an exit chute by the burner and then into cold water. The kiln shown in Fig. 7.3.4.B was developed in Nepal. It uses a steam/kerosene burner, but any forced draft oil or gas burner can be used.

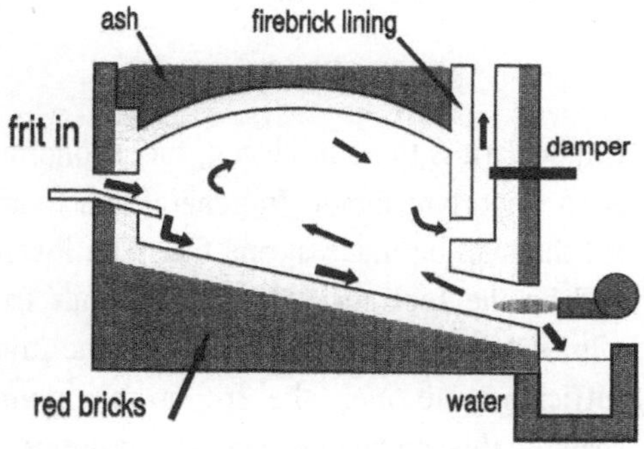

FIGURE 7.3.4.B Side elevation of a continuous flow kiln.

The rate of flow is controlled by introducing limited amounts of raw frit. Too much frit at one time may result in incomplete melting. If the frit runs very fast through the kiln, the low melting materials will not melt properly together with the silica. This may be a cause of water-soluble frit (see page 48).

The frit can be slowed down in the kiln by making less of a slope and by putting some obstacles in the way (like kiln shelf supports).

7.3.5. ROTARY

Rotary frit kilns are large refractory-lined cylinders, which have a burner (gas or oil) that passes through them. The raw frit is introduced, and the kiln rotates full turns (or back and forth) as the frit melts. This has the double purpose of ensuring good mixing and of transferring the heat of the firebrick lining to the frit as this constantly moves over it. When the frit is completely melted, the kiln is turned so that the frit flows out through an opening into cold water.

FIGURE 7.3.5.A Front and side elevation of a rotary frit kiln. It consists of a firebrick-lined steel drum resting on rollers. It is gas- or oil-fired.

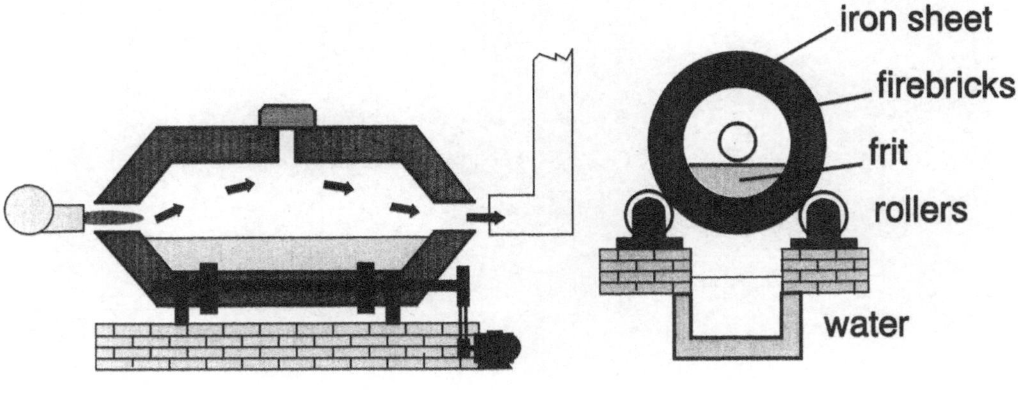

7.3.6. FUEL ECONOMY

If much frit is to be produced, fuel economy is an important factor. In general, the more frit that can be made at one time, the lower will be the fuel cost. In a continuous frit kiln, it takes several hours to heat the kiln sufficiently to melt the frit at maximum speed – this preheating period consumes a lot of fuel. It is best to fire several hundred kg of frit at the same time to reduce firing costs.

Frit industries generally use rotary kilns, as they are the most economical for long, continuous use. However, the continuous kiln developed in Nepal by the Ceramics Promotion Project compares favorably with standard fuel/frit ratios obtained with rotary furnaces.

Examples of fuel to melted frit ratios are:

Frit kiln type	Batch amount	Fritting time	kcal/kg melted frit
Open hearth coal	1 – 2 tonnes	6 – 12 hours	7500 – 11250
Nepal, continuous flow	1.5 – 2 tonnes	48 hours	5150
Rotary, India	300 kg	2 hours	5700

FIGURE 7.3.6. A rotary frit kiln. Enamalgar, Calcutta, India.

8. PREPARATION OF GLAZES

Glazes should be prepared in a systematic manner in order to prevent mistakes. Most problems with glazes come from simple things, like incorrect weighing, mistakes in identifying raw materials or not sieving the glaze correctly.

Glaze mistakes are expensive, as they can result in the loss of an entire kilnload. For this reason, it is important to have the right person in charge of making glazes – cleanliness, orderliness, careful record-keeping, and reliability are required.

Most small producers do not need a large variety of glazes – in fact, many use only one or two standard glazes and achieve va-

FIGURE 8.0.0.A Packing of base glaze and "masala" colorants in the shop of a small glaze supplier in Nepal. The small colourant bags are added to a large bag of base glaze producing a standard color.

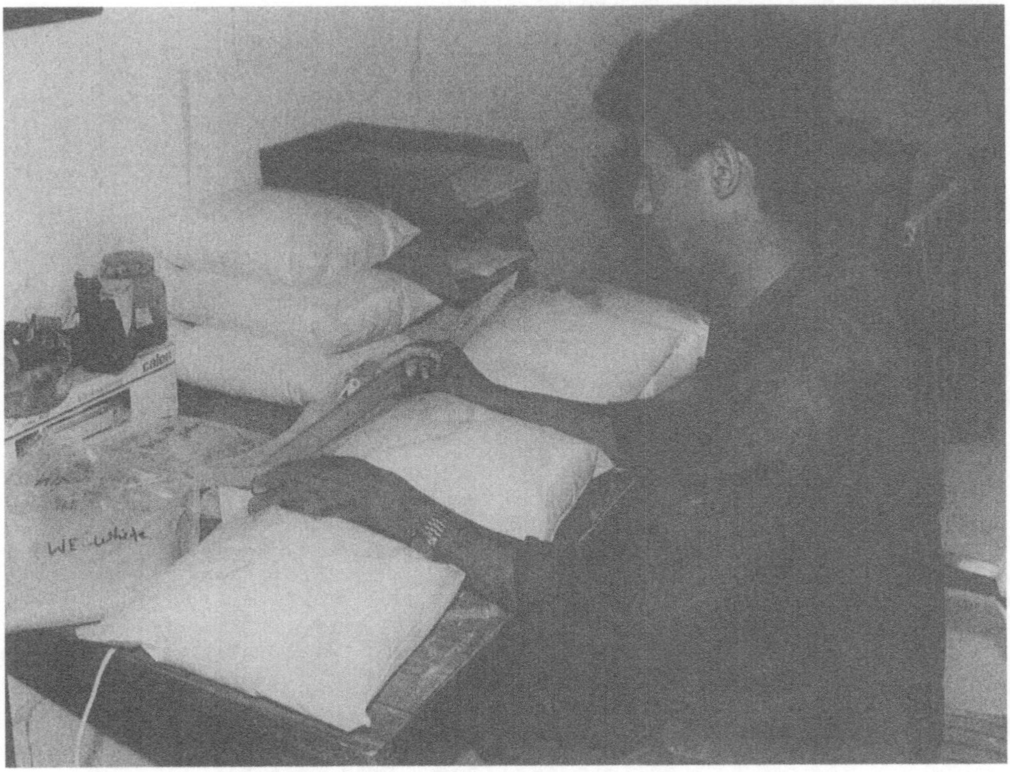

riety by changing the colors, doubleglazing or using engobe decoration.

Designing a glaze is somewhat like choosing a paint in the paint store. First of all, you must decide if you want a glossy or matt surface, transparent or opaque. Then you can add different colors.

Base glaze

The base glaze is simply the combination of materials that melts at the desired temperature. It is either transparent or opaque, matt, semimatt, glossy etc. without any particular color.

Glaze additions

These are usually coloring oxides that are added to the glaze. In Nepal a glaze supplying system serving small producers was established. A base glaze was supplied in 5-kg bags and 8 different colors were supplied in small bags that produced standard colors when added to 5 kg base glaze. The small bags contain coloring oxides mixed with a small amount of base glaze so the colors disperse more easily in the base glaze.

8.1. RAW MATERIALS REQUIREMENTS

Raw materials need to be as reliable as possible and always ground to the same mesh. If obtained from a glaze supplier, the materials are usually ground to at least 100 mesh. Because materials that are finely ground melt more easily, some ingredients may be as fine as 400 mesh. This is particularly true of quartz – 200-mesh quartz will produce a different result than 400-mesh quartz.

When you get new raw materials, they always should be tested before using them in production. The best way is to try them in a standard glaze that you know well and to compare the results with the known glaze.

8.2. GRINDING GLAZE MATERIALS

8.2.1. COARSE MATERIALS

There are several steps in grinding glaze materials. Since many of them (feldspar, quartz, limestone) come as rocks, they first need to be reduced to pebble size. Feldspar and quartz rocks are first calcined to make

FIGURE 8.2.1.A Simple updraft kiln for calcining quartz and feldspar.

them soft enough to crush. Calcining means firing to just above 600°C. This can be done in the cold spots of a biscuit firing or for large productions in a special kiln. Crushing of small amounts can be done with a hammer (use eye protection), and large amounts are usually done in a jaw crusher.

8.2.2. BALL MILLING

Ball mill operation

Ball mills are used for fine grinding of ceramic materials. The material has to be reduced to sand size (2 mm or less) before grinding in a ball mill.

Some typical uses of ball mills are:

- grinding clay that does not easily slake
- preparation of casting slips
- grinding of body additions like feldspar, quartz and glass powder
- grinding of frit granules
- grinding of glazes
- grinding of engobes and terra sigillata
- preparing color pigments for glaze, engobes or bodies.

There are two main types of mills:

- Large mills with an axle system are called *ball mills.*
- Small mills are called *pot mills* or *jar mills.*

FIGURE 8.2.2.A Ball mill with axle system.

FIGURE 8.2.2.B Pot mill with two porcelain jars in operation.

These are usually small (up to 5 liter) porcelain jars or plastic jars, which rotate on two rubber-covered rollers.

Conical ball mill

For large production conical ball mills are used. Various sizes of pebbles are used and the material is fed from one end and discharged at the other. Variation of the centrifugal force caused by a conical 30° slope at the discharge side classifies both pebbles and material so only fine material is discharged.

Vibro energy mill

This is a new type of grinding machine consisting of cylindric grinding chamber suspended on springs and vibrated at high frequency with the help of an excentric mounted on an electric motor. The chamber

FIGURE 8.2.2.C Conical ball mills seen from the discharge end. (Courtesy of Co-Bam Ltd.)

FIGURE 8.2.2.D Vibro energy mill. (Courtesy of William Boulton Vibro Energy Ltd.)

is completely packed with very hard small cylinders between which the material is filled. The vibrations make the small cylinders grind against each other and the material to be ground. The vibrating mill is better at ultrafine grinding and is more energy-efficient than ball mills.

Lining

The grinding action takes place between the pebbles, and *not* between the pebbles and lining. Therefore a ball mill with a steel drum can work without a lining (except for white body, where rust particles will cause discoloration). Pebbles constantly falling on a steel drum make a lot of noise. A lining will reduce the noise and at the same time prolong the life of the steel. Traditionally, linings are made of porcelain or stoneware bricks set in a cement mortar, using high alumina cement and coarse silica sand. Common cement can be used if necessary but may cause pinhole problems in glazes. The bricks should be dense and vitreous. A porcelain body for lining bricks and pebbles (fire to 1250°C or higher) is:

China clay	40 %
Quartz	25 %
Feldspar	30 %
Bond clay	5 %

One type of brick is made concave to fit the curve of the drum and another type is made for the end walls of the drum.

Linings can be made from granite, quartzite or similar hard rocks (not limestone or marble). They are cut to shape and set in a high alumina cement mortar. They last far longer than porcelain bricks. Stoneware bricks can be used for the end walls, which are worn out more slowly.

Instead of a hard lining, thick rubber sheet glued to the inside makes a very long-lasting and quiet lining.

Pebbles

Pebbles or balls can be made from vitreous clay bodies. However, it is often cheaper to collect stones of granite, quartz or quartzite along riverbeds. Flint, a variety of quartz, is excellent for pebbles. The hardness is tested with a penknife to make sure it is above 5.5 (see Mohs' scale page 41). Pebbles of limestone are not satisfactory, as they contaminate the glaze. The shape should not be flat or elongated but spherical. (Cylinders of equal diameter and length are sometimes used to obtain particles with less variation in particle size.) Size should be between 2.5 and 5 cm in diameter.

Pebbles wear out, so occasionally take out all the pebbles for inspection. Those that are broken or flat should be discarded. In large mills, pebbles are removed when they are less than 2-3 cm. In small mills pebbles smaller than 1.5-2 cm are discarded.

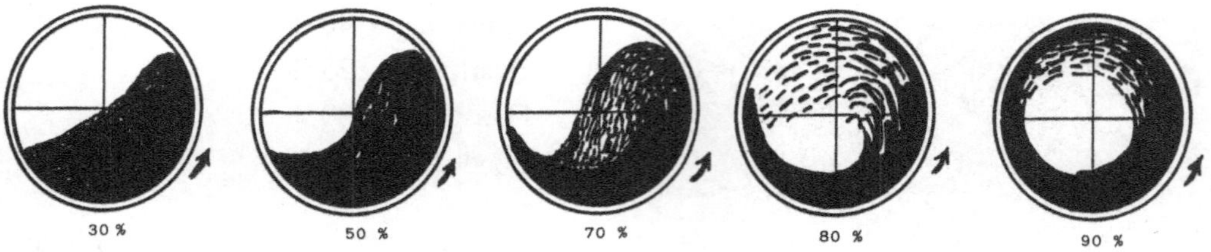

FIGURE 8.2.2.E A cross section of a ball mill running at speeds from 30-90% of critical speed. At 30% the grinding takes place mainly between pebbles and lining, at 70% a good cascading rolling produces efficient grinding, and at 90% very little grinding takes place.

As the pebbles grind down, they contribute a small amount of material to the glaze. Usually this is not enough to make a difference in the glaze. However, if you have glaze problems that cannot be traced to any other cause, the ball mill pebbles should be checked.

Ball mill speed

Grinding of material takes place between the pebbles of the ball mill as they roll down the slope of the cylinder. If the speed is too high, the grinding action stops because centrifugal force stops the pebbles from falling.

This happens when the cylinder is running at its critical speed. Critical speed is calculated from the inside diameter of the cylinder:

$$\text{Critical speed in rpm} = \frac{29.9}{\sqrt{r}}$$

(r = internal radius in meters)

The actual speed of the ball mill should be 60 - 80 % of critical speed. Small ball mills can be closer to 80 % and large ones closer to 60 % Appropriate speed can be read from Fig. 8.2.2.F.

The most efficient grinding is achieved when the pebbles roll as shown in the center ball mill of Fig. 8.2.2.E. The pebbles cascade in a steady stream, and grinding takes place between the pebbles. The speed of the ball mill at 80% is too high. The pebbles have started to fall freely and this causes excessive wear as the pebbles hit one another and the lining.

Unfortunately it is not possible to look inside during milling, but if the pebbles make a low, rumbling sound the speed is correct. If they make a loud banging noise, the speed

FIGURE 8.2.2.F Graph of ball mill speeds.

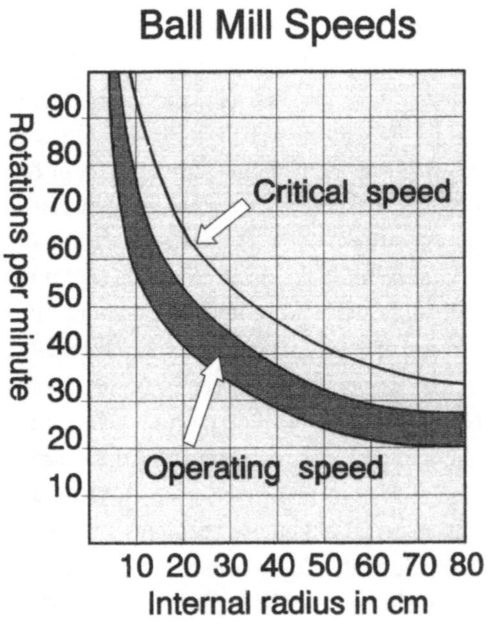

is too high or there is too much water, charge or pebbles in the mill. Porcelain jar mills crack if they run at too high a speed.

Charge

With a speed of 60 - 80% of critical speed the charge should be:
(by volume)

Pebbles:	45 - 55 %
Water:	12 - 20 %
Material:	20 - 25 %

When the mill is filled to maximum capacity, the speed should be closer to 60% of critical speed. The water content should be enough to produce a thin slip. After filling, about 30 % of the volume should remain empty. If you measure all the materials separately, total volume may seem to be 85% of ball mill capacity. However, since the water and material fill the spaces between the balls, this will still result in 30% empty space.

Example

A ball mill with new lining measures inside:

width 0.64 m, diameter 0.445 m
volume of ball mill $= \pi \times r^2 \times w$ or 320 liters
critical speed $= \frac{29.9}{\sqrt{r}}$ or 52.8 rpm

60%-80%
of critical speed = 31.7 rpm – 42.3 rpm

Charge:
(by volume)

Pebbles	144-176 l.
Water	38-64 l.
Material	64-80 l.

A typical glaze has a density (specific gravity) of approximately 2.7. That means that the glaze charge should be 24-30 kg.

Ball milling time

The time for ball milling varies with the hardness of materials. Soft materials such as frits may require only 2-3 hours, whereas hard materials like quartz can take 24 hours or more.

When you ball-mill standard materials, it is important to mill each batch for the same amount of time. For this reason, it is a wise investment to purchase a timer switch for the mill. This will avoid human errors. Too much ball milling can cause glaze crawling.

Operating procedure

Before each operation:

1) Check that the ball mill is clean inside.
2) Check that pebbles fill half of the ball mill – refill if necessary.
3) Fill in materials (20-25% of mill volume).
4) Fill water until pebbles and material are just covered.
5) Be *very careful* about correct ball milling time. If possible, use an automatic timer.

After operation:

6) After emptying the ball mill, clean it thoroughly with water by filling it and running it with the pebbles. If the same material is to be ground, cleaning is not needed.

Every month:

7) Empty the pebbles out and remove all pebbles that are too flat or less than 2 cm in diameter.
8) Inspect the inside lining for signs of wear, and repair as necessary.

8.3. WEIGHING, MIXING, USING BATCH CARDS

Weighing glaze ingredients

First, you must have an accurate scale. This can be a small balance, such as is used by jewelers, or a triple beam balance, which is faster to use. Spring scales are not accurate enough, nor are postal scales. For large quantities, the most accurate low-cost balance is the common beam balance which uses standard weights.

FIGURE 8.3.0.B Example of a glaze batch card used for quality control.

Batch cards

For best results, a batch card system should be used. These are simply cards that have the glaze recipe written on them. As each ingredient is weighed, it is checked off on the list. When all materials are weighed, the batch card is given a number (usually the date). The same number is written on the glaze container. This makes it easier to find out the problem when the glaze does not work correctly.

FIGURE 8.3.0.A Weighing glaze materials.

Water
The ingredients are then added to a container which already has the approximate amount of water in it. CAUTION: The water must always be clean. After mixing, the water is adjusted. **It is always best to start with less water than required**. If the glaze is too fluid, it is difficult to remove excess water.

Containers
Containers for glaze should be plastic or wooden. Metal containers cause contamination and rapid settling of glazes.

8.4. SIEVING

Glazes are normally sieved through a 100-mesh screen. The glaze should be poured through without forcing it. **Never use your hand to force glaze through a sieve**, as this will quickly break down the wire mesh. A brush should be used instead.

8.5. SUSPENDING AND BINDING AGENTS

Because glazes are mixtures and not solutions, they tend to settle at the bottom of the container. Normally, the clay content of the glaze will be sufficient to keep them in suspension during application. However, some glazes tend to settle as a cement-like layer on the bottom and are difficult to stir. These glazes require the addition of a suspending agent.

FIGURE 8.4.0.A Sieving glaze with the help of a brush.

Suspending agent

The most common suspending agent is bentonite, in 1-2% additions. This will normally not be enough to affect the glaze when fired. Dry bentonite cannot be added to wet glaze, as it will just form lumps and be impossible to mix in thoroughly. Instead it should either be mixed separately with water into a thin slip and then added to the glaze or it should be added to the dry glaze and mixed in well before adding water.

Binder

Another common problem is that some glazes tend to be powdery, and come off when loading the kiln. For this problem a binder is added.

Bentonite also works as a binder and is the simplest to use. Another common binder is CMC gum (carboxymethyl cellulose), which is available in either liquid or powder form. The liquid can be used directly, about 1%. The powder needs to be dissolved in water (1:10) overnight and then is added to the glaze as liquid.

Organic binders such as gum arabic, wheat flour, sugar or starch (0.1-0.5% of dry glaze) are sometimes used. These have the disadvantage of fermenting. They should be used immediately after mixing, or if stored a few drops of chlorine bleach or formaldehyde can be added as a preservative.

Addition of 1% raw borax produces a hard surface that does not powder when painted on.

Flocculation

Addition of a flocculation agent will make the glaze more creamy. The pottery will absorb the water more easily so glaze is picked up faster.

This works better in combination with clay or bentonite. Common flocculants are: Epsom salts (magnesium sulfate), calcium chloride, calcium nitrate and borax. They are prepared by adding 100 g flocculant to 200 ml hot water and the solution is added to the glaze one tablespoonful at a time (up to 1% of dry glaze weight). Plaster of paris (already set) can also be used.

Flocculation is also used for nonporous ware often in combination with a binder. The creamy glaze forms a thick loose layer that stays on the nonporous surface.

Deflocculation

When the glaze is deflocculated it becomes more fluid with the same amount of water. This is sometimes used for glazing nonporous ware that cannot absorb water. Sodium silicate and soda ash are the most common deflocculants and they are prepared in the same way as flocculants.

CAUTION: Binders, deflocculants or flocculants should only be added after the glaze is ball-milled.

8.6. DENSITY (SPECIFIC GRAVITY)

Most potters judge the consistency of their glaze by experience and feel, or by test application to a few pieces of biscuit to see if the thickness is correct. The standard test is to check thickness with a fingernail, which is a very accurate test for an experienced glazer. Then adjust the water as necessary.

A more accurate method is to measure the specific gravity of the glaze with a hydrometer, such as is commonly used to

FIGURE 8.6.0.A Dipping of dinner plates at Royal Copenhagen is still done manually. The company found that a very sophisticated new glazing machine could not do it as well as an experienced glazer.

FIGURE 8.6.0.B Hydrometer made from a glass test tube.

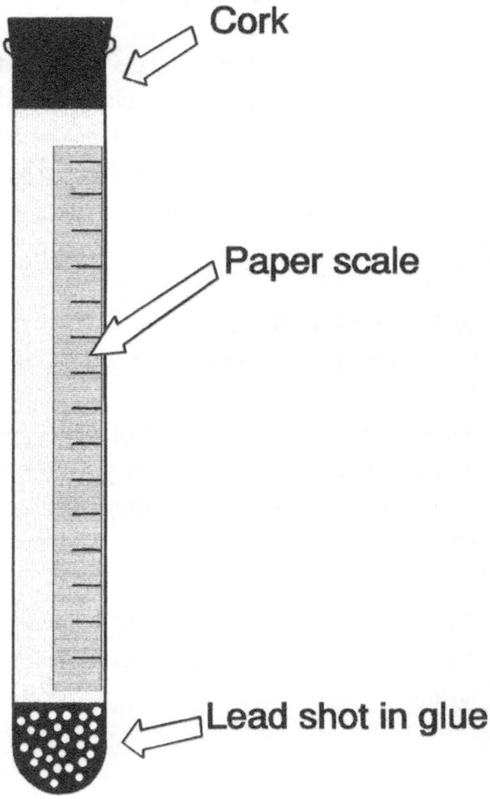

judge the amount of water that has been mixed with milk. When reading the depth the hydrometer sinks, take care that it is really showing the correct density. If the glaze is thick you have to vibrate the bucket repeatedly to make sure the hydrometer sinks in.

Specific gravity (s.g.) is a measure of the density of a liquid compared to water, which has a standard specific gravity of 1. Glazes will always be heavier than water. The specific gravity is found by weighing a specific volume, say 1000 ml (milliliters). If this weighs 1500 g the s.g. is 1.5. Weighing is more accurate than using a hydrometer.

After you find out the correct amount of water by trial and error, the specific gravity can be measured and future batches of the same glaze made to the same specific gravity.

FIGURE 8.6.0.C Glaze density is measured with a homemade hydrometer.

CAUTION: This is not always a reliable method because the water absorption of your biscuit will vary with its firing temperature. The water will still need to be adjusted by trial and error. Trial application and testing with a fingernail still constitute the most reliable method.

8.7. OLD GLAZES, PROBLEMS

If you keep wet glazes around for a long time, they will usually have problems with settling or drying up. These glazes can still be used but it will be necessary to adjust the water and to resieve them. If the glaze is extremely thick, it is sometimes best to dry it out completely, crush it and remix it.

Before using a glaze that has set in the bucket for a few days, it should always be sieved through 60 or 100 mesh.

FIGURE 8.7.0.A Crystals formed in a frit glaze. The frit was soluble owing to lack of quartz in its recipe.

Too much water in the glaze is also a problem. The glaze can be allowed to settle and excess water carefully taken off the top. CAUTION: With soluble glazes, this can remove some of the ingredients and result in a glaze that no longer works correctly. In this case, the water should be allowed to evaporate until the thickness is correct.

Glazes made with raw borax, or incomplete borax frits, will often grow crystals. These cannot be sieved. The glaze should be dried out, the crystals crushed and remixed.

Some glazes will develop mold and begin to smell. Although they can still be used, it is probably better to just throw them out.

8.8. TEST YOUR GLAZES!

The wise potter will never glaze a kilnload with untested glaze. Enough glaze should be kept on hand, so that each new batch can be test-fired in the regular glaze firing before it is used for application.

8.9. COMMERCIAL PRODUCTION OF GLAZES

Glazes that are sold commercially are usually in dry powder form. They are made as standard glazes by ball milling, then are dried and packaged.

These glazes are simply mixed with the correct amount of water and sieved before using.

9. GLAZE APPLICATION

Glaze application is a skill that takes some time to learn. In order to get consistent results, it needs to be done carefully and the same way every time. Thin and thick application will give different results, and careless application is always ruinous.

Glazing should be done just before loading the kiln, as glazed pieces that lie around gather dust and get damaged. Some glazes tend to crawl if fired right after glazing. If you have such problems, allow the glazed ware time to dry completely before firing.

9.1. WORK PLACE, CLEANING AREA

Before glazing, you should have a neat and clean area to work in. Dust thoroughly and remove small children. The biscuit to be glazed should be organized in one place, with all like items grouped together (cups, bowls, vases etc.). Ware boards are cleaned and arranged, ready to take the glazed ware to the kiln. The glaze should be sieved and checked just before starting the application. Clean water and sponges should be available.

Large items are usually glazed first, as they require a full bucket for even application.

Correct application depends on many different factors:

- Density of the glaze
- Viscosity of the glaze
- Particle size (depending on grinding time)
- Expertise of the worker
- Porosity of the biscuit
- Thickness of the piece
- Dipping time.

Although some of these factors can be controlled accurately in large industries, the small producer will have to depend on experience. Mistakes will be made at first, and it is important to be able to understand what went wrong, so it can be corrected.

9.2. APPLICATION METHODS

The particular method of applying glaze depends on the type of ware – small, big, sculpture, tiles, open forms, closed forms etc.

Generally the inside of an object is glazed before the outside, to prevent handling defects.

Loading systems need to be considered carefully. Most pots are loaded on shelves directly, so the feet must be left unglazed. If foot rings are to be glazed, then each piece must be individually set on special kiln furniture in the kiln.

9.2.1. PAINTING

Glaze is sometimes applied with a brush. This is not recommended because it takes a great deal of skill to obtain an even coat, as well as a lot of time. Painting is used on sculptural objects that cannot be dipped or sprayed. Three to four coats are brushed on, letting each coat dry before applying the next. In order to see each coat, sometimes organic color dye (food coloring) is added.

9.2.2. DIPPING AND POURING

Dipping and pouring constitute the most common method.

The glaze needs to be stirred frequently during application time.

FIGURE 9.2.2.A Glazing a cup.

1) Fill the inside of the cup with glaze.

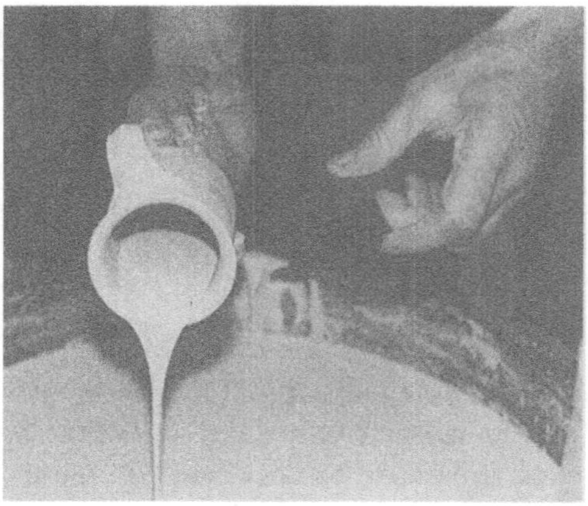

2) Rotate the cup while at the same time pouring out the glaze.

3) Holding the cup by the foot, dip it in the glaze.

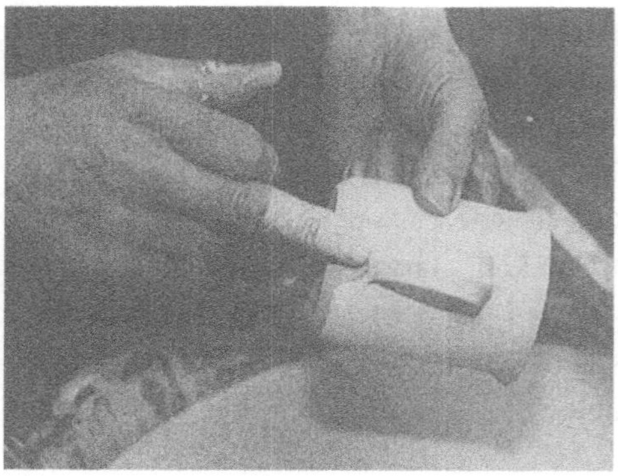

4) Unglazed spots near the foot can be covered with glaze on a finger.

FIGURE 9.2.2.B Glazing a cup in one dip with the help of a pair of tongs.

FIGURE 9.2.2.C Large items that cannot be dipped are glazed by pouring from a pitcher while held over the glaze bucket.

FIGURE 9.2.2.D Three steps of glazing the inside and outside of a cup in one dip.

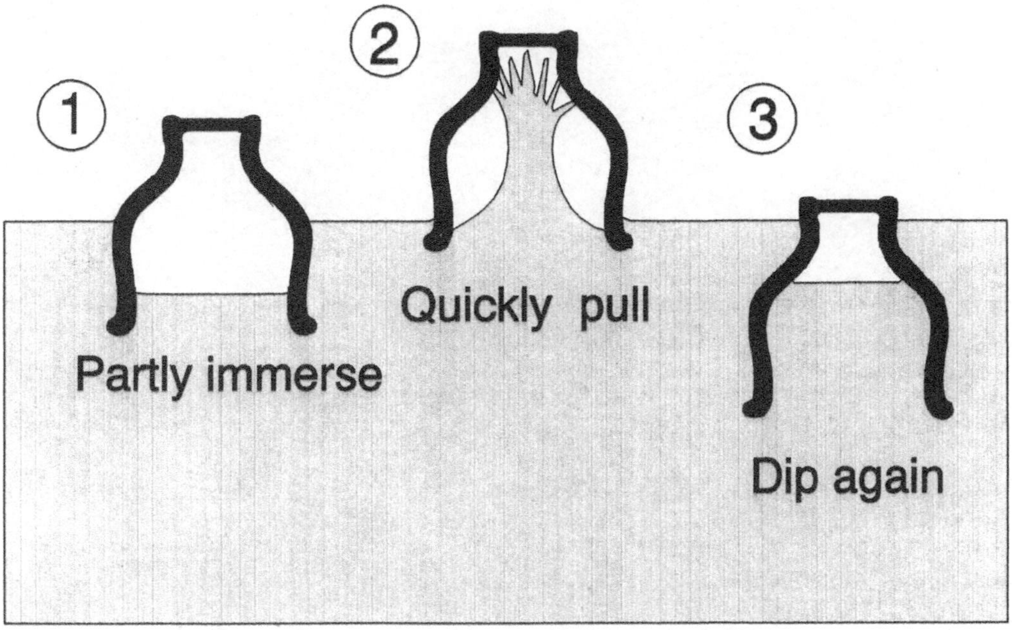

Cups and bowls

Cups can be glazed inside and out in one movement (after some practice). Hold the cup by the foot and dip it at a slant to let glaze inside, while the outside is also coated with glaze. Then quickly pull up and push down. This results in a "fountain" of glaze that covers the entire inside.

FIGURE 9.2.2.E Tongs for glaze dipping made by twisting three wires to fit the fingers.

FIGURE 9.2.2.F A bowl is glazed inside by pouring glaze inside the spinning bowl.

FIGURE 9.2.2.G The bowls are glazed outside using a suction disc to hold the bowls. Lampang, Thailand.

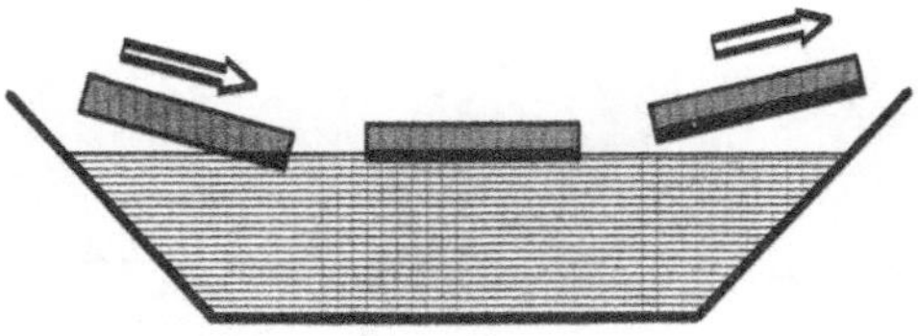

FIGURE 9.2.2.H Dipping tiles in glaze.

Tiles

To dip tiles, hold them by the edges and dip them in the glaze while moving sideways. This also requires practice!

Double dipping

Applying a second coat of the same or a different glaze over the first is known as double dipping. This often happens inadvertently. When glazing the inside, sometimes there will be runs of glaze on the outside. These should be sponged clean before doing the outside. Larger items are often partly dipped to cover the top, then turned over and dipped again to coat the bottom. This usually results in a line of double glaze, which will look different. If the overlapping area is chosen carefully, it can become a part of the design. Otherwise, it will look like a mistake.

For decorative effects, a pot is sometimes dipped partly in one glaze and then again in a different glaze. This results in a third color where the two overlap.

FIGURE 9.2.2.I Bottoms of mugs are glazed first, then the foot is scraped on wire mesh and a sponge. Finally the top part is dipped in another glaze partly over the bottom layer for effect. Khurja, India.

Waterfall glazing

In the commercial glazing of tiles, the "waterfall" system is used. This consists of a conveyor belt, which carries the tiles under a thin waterfall of glaze that pours over them. The thickness of application is controlled by the speed of the conveyor belt and the amount of glaze flow. Excess glaze runs into a tank, which is again pumped up to the waterfall. These machines are often equipped with automatic cleaners that take excess glaze off the sides of the tiles.

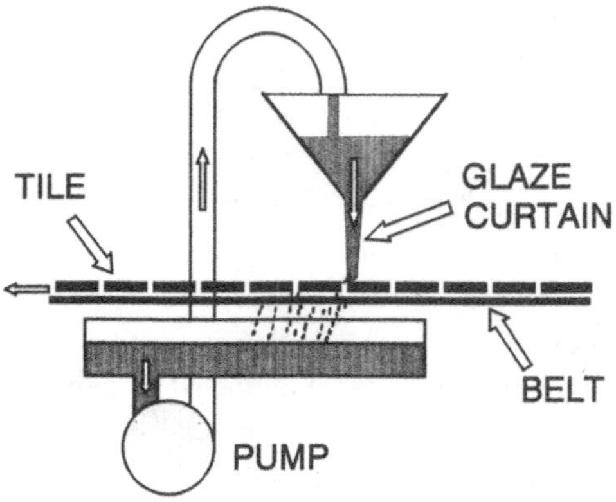

FIGURE 9.2.2.J Waterfall glazing of tiles. The tiles run through a curtain of glaze which is continuously recycled with the help of a pump.

9.2.3. SPRAYING

Spraying is used for items that cannot easily be dipped or poured. It requires an air compressor and a spray gun, as well as a spray booth equipped with an exhaust fan. This is not recommended for the small producer, unless it is required for frequent use or for

FIGURE 9.2.3.A Automatic spraying of plates (courtesy of Royal Copenhagen).

special decorative effects. Ordinary spray guns for paint can be used, but they wear out quickly because glaze is abrasive. Special spray guns for glaze are equipped with silicon carbide spray heads.

Spraying has the disadvantage of wasting a lot of glaze that goes into the air. This is dangerous to inhale, and a spray booth should be provided with an exhaust fan to the outside, as well as having a filter to catch excess glaze. If a great deal of spraying is done, the excess glaze can be collected from the filter and the inside of the booth and reused.

As usual, the inside of the item is glazed first (usually by pouring), and the spraying is done in several even, systematic coats. Each one must be applied before the first one dries, or the glaze may lift off the pot. Each coat should be lightly applied, so that it looks a bit powdery.

It is difficult to judge the correct thickness of glaze and to get it even all over, especially in difficult areas like under handles. In time the glazer will learn to measure the thickness by feeling it with a fingernail.

FIGURE 9.2.3.B Tiles are glazed in a spray booth with an exhaust fan. Compressed air for the spray gun is supplied by a compressor. The operator wears a dust mask for protection. Yangon, Myanmar.

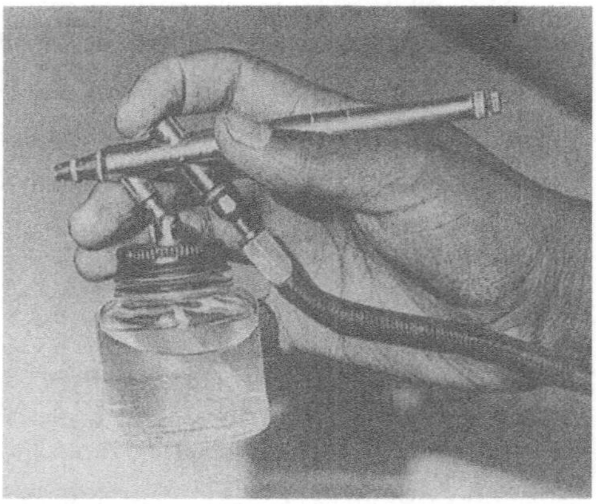

FIGURE 9.2.3.C Airbrush used for decoration.

Airbrush

An airbrush is a very small spray gun that can be adjusted from a pencil-thin spray to a wide pattern. These are not used for glaze application, but are often used for decorative effects with underglazes and overglazes.

Care of the spray gun

Spray guns are very sensitive. They tend to get clogged, so make sure that your glaze is sieved before putting it in the gun. Clean the spray gun immediately after use by rinsing it out and spraying clean water through it until there is no sign of glaze. Glaze left in the spray gun will corrode it and make it unusable.

Glaze fountain

For glazing the inside of large items a glaze fountain as shown in Fig. 9.2.3.D is helpful. The pot is placed over a nozzle from which an electric pump provides a powerful upward shower of glaze when activated with a switch on the floor.

FIGURE 9.2.3.D Glaze fountain for glazing the inside of large pots.

FIGURE 9.2.3.E The bottoms of pots are cleaned of glaze with the help of a continuous sponge.

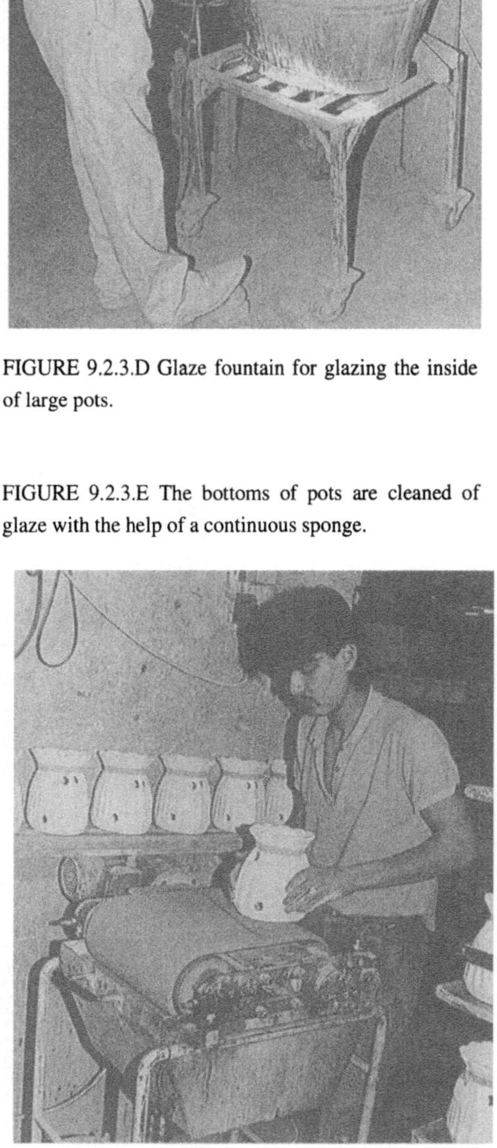

9.3. DENSITY, BINDERS, GLAZE THICKNESS

As described above, it is important to have the correct amount of water in your glaze. The glaze should always be checked and corrected by test dipping some biscuit before starting and then relying on your experience to judge if the thickness is correct. Checking specific gravity with a hydrometer or by weighing is a good practice but should not be relied on.

It is best not to use binders unless you have no choice. CMC gum is the most satisfactory (page 62).

Nonporous biscuit

As previously mentioned, differences in biscuit firing temperature cause differences in porosity and can cause problems in glaze application. Overfired biscuit is especially difficult to glaze, as it will not absorb water. In the making of whiteware, the biscuit temperature is usually higher than the glaze temperature. This results in a semivitrified body that has special glaze application problems. If it is necessary to reglaze pots that have firing defects, they also require special handling.

If you only have a few pieces, they can be heated until almost too hot to handle and then dipped, poured or sprayed (spraying is most satisfactory). The heat will make excess water evaporate.

If glazing vitrified ware is part of your standard production, then it is best to flocculate your glaze (see page 62). This is the opposite of deflocculation (as used with casting slip) and results in a thick, pudding-like glaze with the normal water content.

9.4. WAXING

In order to keep glaze from being applied to the foot of your pots, it is often more efficient to wax the bottoms as compared to sponging them clean. The coating of wax prevents glaze from sticking. There are two common waxing methods:

Hot wax

Paraffin wax is kept melted in a shallow metal pan over an electric heater or a smoldering charcoal fire (an open fire should not be used as the paraffin may start to burn). It should be hot, but not so hot that it starts to smoke. Before applying the glaze, the foot rings are dipped in the paraffin.

Liquid wax resist

It is much easier to use liquid wax resist, which is a wax emulsion in a water base. It can be thinned with water but after drying cannot be dissolved. This is commercially available in some countries specifically for

FIGURE 9.4.0.A Wax is applied to the bottom of a stoneware bowl. The bowl is lightly stamped on a piece of cotton soaked in wax. Myanmar.

glaze application. It is also possible to use liquid floor wax.

Liquid wax resist is also used for decoration.

9.5. SINGLE-FIRE GLAZING

Single-fire glazing is sometimes called "raw glazing", but this term is confusing as "raw glaze" also is used for unfritted lead or borax glazes. Glaze is applied directly to bone-dry or leather-hard ware and fired once up to the glaze temperature. Not all glazes and bodies are suitable for single firing, and each combination needs to be tested.

Glazes that work on biscuit ware will often also work on bone-dry clay with a small addition of a plastic clay or bentonite. Glazes for leather-hard glazing will need more clay so the glaze layer will shrink along with the clay during drying. The leather-hard method is less practical, since each batch of leather-hard pots must be glazed immediately, causing problems in the work flow.

The advantage of single firing is that it avoids the fuel and extra handling needed for biscuit firing. The main problem with single firing is crawling caused by different shrinkage rates of clay and glaze in the early stages of the firing. Single-fire glazes usually have a high percentage of clay.

Delicate ware cannot usually be single-fired successfully, as it tends to be damaged by the water.

Single-fire glazing needs to be done quickly and carefully, without letting glaze stand inside the pot for a long time. Dipping and pouring can be used, and spraying is also effective.

Firing needs to be done more slowly than usual, so that pots do not explode. The early stages of firing should be done as with biscuit firing.

Single firing is used most often in large tile industries, where it saves fuel.

FIGURE 9.6.0.A A kiln loaded with stoneware.

9.6. HANDLING, DRYING BEFORE FIRING

Good glaze application requires careful handling. Many pots are spoiled by fingerprints or glaze that is knocked off during handling. Pots should be allowed to dry before loading in the kiln.

The kiln loader should be responsible for checking each pot as he places it in the kiln. This means inspecting the foot to see if it is clean and rejecting pots with damaged or thick glaze. The loader should constantly clean his hands of glaze dust especially when loading ware with different colored glazes. Otherwise colored fingerprints will mark the pots.

FIGURE 9.7.0.A Salt is thrown into the firebox of a coal-fired sewage pipe kiln. Butwal, Nepal.

9.7. SALT GLAZING

In salt glazing, no glaze is actually applied to the pot before firing. The ware is single-fired up to the maturing point of the clay and rock salt is then introduced directly into the firebox. The salt breaks down into sodium and chlorine gas. The sodium combines with silica on the surface of the pot to make a durable glaze and the chlorine goes up the chimney, combining with water in the air to form hydrochloric acid. This is an irritant, as well as causing damage to vegetation and metal structures in the immediate vicinity. Another problem is that the salt erodes the firebricks in the kiln rather fast.

Salt glazing normally is done on stoneware at temperatures above 1100°C. Salt is often mixed with borax to lower the melting point (see also page 19).

10. DECORATION

Decoration is a very big field, which deserves a separate book to cover it in detail. Here we will only discuss some of the main techniques for using glazes and engobes.

10.1. DECORATION AND DESIGN

The main reason for decorating pots is for pure enjoyment. As pottery is something that is used intimately every day, it should be attractive and interesting, besides being simply functional. Decorated pottery also has a better market value and often more than pays for the extra time taken.

Good decoration is always related to the design of the pot. It should be used to emphasize and enhance the shape of the pot, rather than being applied randomly.

There are several approaches to decoration:

Banding

Plain or decorative bands of color are painted around the pot, usually by spinning the pot on a banding wheel and applying color with a brush. The bands are placed where they emphasize changes in the curve of the pot, for example, at the rim, the belly, the shoulder.

FIGURE 10.1.0.A Example of banding.

FIGURE 10.1.0.B Example of area decoration. Aa.Kvorning.

Area decoration

Decoration is placed inside a defined area, such as a circle. Again this should be done to emphasize the natural curves of the pot.

Overall patterns

These are patterns that are repeated around the pot, often expanding and contracting as the pot does.

Contrasting shapes

These are strongly shaped areas of pattern or color that contrast with the shape of the pot.

FIGURE 10.1.0.D The square shape of the plate contrasts with the circular decoration. Aa.Kvorning.

FIGURE 10.1.0.C Example of overall pattern decoration. Portugal.

10.1.1. MOTIFS, STYLES, LOCAL INPUTS

There are as many motifs and styles of decoration as there are cultures in the world. In traditional cultures, motifs are selected from mythology and familiar designs. Nowadays, with the mixing of cultures around the world, pots are often designed for what can be marketed for export, and design tends to be based on fashion rather than tradition.

The potter selling to tourists will generally choose traditional motifs, since tourists are interested in the culture of the area.

FIGURE 10.1.1.A A harvesting scene. A local tradition of decoration has been transferred to engobe technique by Maithili women from Janakhpur, Nepal.

10.2. GLAZE DECORATION

Glaze decoration is done with the glaze itself or with colorants under the glaze or on top of the glaze.

10.2.1. UNDERGLAZE

Underglaze decoration is decoration that is applied under the glaze. It is affected by the transparency and fluidity of the glaze.

Underglazing is usually done under a transparent glaze in order to show it clearly. However, beautiful effects can be obtained under opaque or semiopaque glazes.

A variety of pigments and oxides may be used.

Metallic oxides
The more fusible metallic oxides can be used directly as underglaze pigments, mixed

thinly with water. The satisfactory ones are red iron oxide, cobalt carbonate, manganese dioxide and copper carbonate. Designs made with oxides alone will often run with the glaze. Refractory oxides, such as chrome oxide and rutile, can cause crawling.

Oxides mixed with glaze

Metallic oxides can be mixed about 50/50 with glaze, which will prevent the problem of crawling. However, the decoration will usually flow with the glaze and should be designed with this in mind.

Underglaze pigments

These are pigments that are specially prepared by fritting metallic oxides in a base glaze that fires hard but is not fluid. Rather than preparing them yourself, it is usually better to purchase commercial underglazes from a supplier. These are supplied for various firing temperatures and firing conditions in a wide range of colors. Not all colors can be used under all conditions, and suppliers can usually tell you which are suitable for oxidation and reduction and what type of base glaze will develop the best colors.

FIGURE 10.2.1.A Painting with underglaze colors on a biscuit-fired porcelain body. Note how neatly the work place is organized (courtesy of Royal Copenhagen).

10.2.2. ON-GLAZE

On-glaze decoration is applied on top of the unfired glaze. It may be done with a contrasting color of glaze or with metallic oxides or glaze pigments.

Application is done by brushing or spraying. Even more than underglaze, on-glaze decoration will tend to flow with the glaze. If distinct patterns are desired, a stiff, viscous glaze will give the best results.

FIGURE 10.2.2.A On-glaze decoration with metallic oxides. Bujora, Tanzania.

FIGURE 10.2.2.B A double-glazed ashtray with a dark bottom glaze and a lighter colored, high-surface-tension glaze on top.

Double glazing

Glazes high in surface tension (see page 30) tend to form into small islands on melting. This may cause crawling, but it can also be used as a decorative effect by applying two different glazes on top of each other. The glazes must have different degrees of surface tension. This is achieved by adding clay or talc to one of the glazes. The colors should be contrasting. The best results are obtained with a light-colored glaze at the bottom.

FIGURE 10.2.2.C Double glazing with the high-surface-tension glaze on top. This draws itself into islands leaving the bottom glaze in irregular patterns.

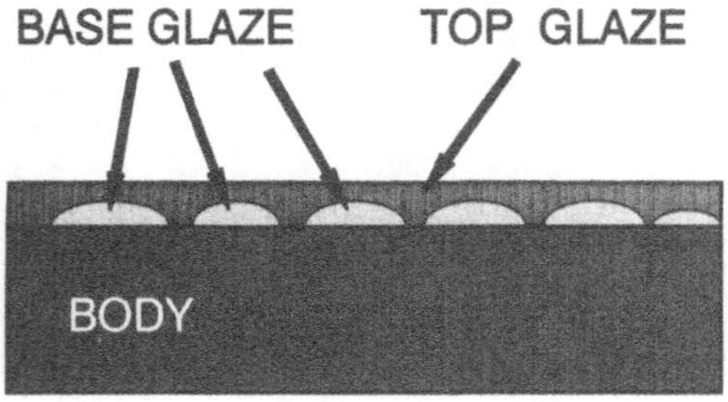

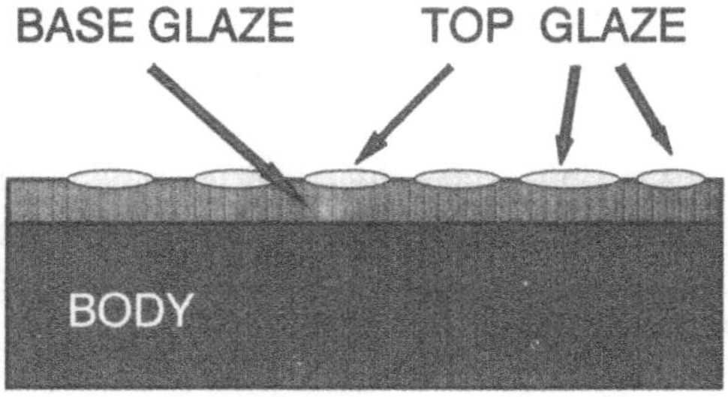

FIGURE 10.2.2.D Double glazing with the low-surface-tension glaze on top. This produces a different effect.

10.2.3. OVERGLAZE

Overglaze decoration is often called "china painting". The pot is glaze-fired as usual and then is decorated with special low-temperature enamels that fire at around 700°C. The enamels are prepared from color pigments and low-temperature frits and are best purchased from commercial suppliers. They are available in every color and have the advantage of firing to true colors, making them suitable for elaborate painting effects. They also stay where they are applied, as there is no chance of the glaze running at this low temperature.

FIGURE 10.2.3.A Decorating with overglaze colors on a porcelain glaze. Singapore.

Overglaze is available as powder, which must be mixed with a medium. This is best done by grinding the pigment and medium on a glass plate with a thin palette knife.

Sometimes plain water is used – this works well when filling areas with solid colors. It helps to add some water-soluble glue (white glue) to provide dry strength.

For more elaborate painting, pigment is mixed with special oils. The best is oil of lavender, which is thickened as desired with a thicker oil. The consistency is controlled very much like with oil paint.

Some suppliers have ready-mixed overglaze, which comes in tubes. This can be used directly, without grinding.

Special metallic or iridescent overglazes are known as "luster". These are available commercially as liquid gold, platinum and a variety of mother of pearl colors. They also are fired at 700°C. NOTE: Lusters take on the same surface as the glaze, i.e. a matt glaze will produce a matt luster and a shiny glaze will give a mirror-like effect.

Overglazes are applied by brushing or by spraying.

Overglaze transfers or decals

Most commercially sold decorated dinnerware is decorated with decals (sometimes called transfers), which are made from overglaze that is silkscreen-printed onto decal paper. These are available from suppliers in a range of standard designs and can also be

FIGURE 10.2.3.B Transfer of decal papers (courtesy of Royal Copenhagen).

custom-made (in large quantities). They are applied to already glaze-fired ware.

The decal is soaked in water until the design can easily be slid off the paper. The wet paper is placed in the correct location and is carefully slid from under the design, leaving the design adhered to the pot. The design is carefully smoothed, dried and fired like standard overglaze.

10.2.4. REGLAZING, MULTIPLE GLAZING

Reglazing means applying glaze and firing an article that already has been fired once. It is sometimes necessary when glazes do not work correctly the first time – they may be too thin, underfired or not the right color.

Multiple glazing is the process of glazing and firing an article two or more times in order to get special glaze effects that cannot be achieved in one firing. It often involves first glaze firing the pot at a high temperature and then glaze firing with lower temperature glazes, in order to get special colors or textures. For example, a pot may be fired to cone 10, then be refired with cone 06 glazes to get bright colors. It may be fired several times at cone 010 for overglazes and lusters.

Reglazing or multiple glazing makes an article more expensive, but it can also be sold at a much higher price.

Reglazing hints

Because already fired ware is no longer porous, it is difficult to apply enough glaze. It helps to first heat the article (as hot as you can hold in your hand), and spraying is the most effective way to apply more glaze.

For multiple glazing, glaze can be specially prepared by adding cellulose gum (CMC). This thickens the glaze and gives it better handling strength.

10.3. ENGOBE DECORATION

Engobe is a specialized type of clay slip that is used for decoration under the glaze. The engobe shows through a transparent or semitransparent glaze and can have the range of color that is possible in glaze.

10.3.1. ADVANTAGE/ DISADVANTAGE COMPARED TO GLAZE DECORATION

Engobes stay where they are applied and do not run with the glaze. This makes it possible to do designs with sharp edges or a lot of detail.

Engobes are often used on dark clay bodies in order to provide a bright, white background for glazes.

10.3.2. ENGOBE MAKING, ADJUSTING TO BODY

Engobe is generally prepared as a white base and then colored with appropriate coloring oxides. If you are already using a white clay body, this becomes an engobe simply by thinning it with water. A dark body will require a white engobe formula that fits it correctly.

The main problem with engobe is getting a good fit between engobe and clay body. It must have about the same amount of shrinkage as the body or it will tend to flake off or crack. The engobe should also mature at the same temperature as the clay body in order to provide a strong clay-glaze interface. Engobes can be applied at three different stages:

Leather-hard

This is the best stage for applying engobes, as it permits the widest range of decorating techniques (brushing, incising, inlaying, stencil etc. – see below). The engobe must have enough clay in it to shrink at the same rate as the body.

Bone-dry

Engobes for bone-dry application need to have less shrinkage, so that they adhere to the body.

Biscuit

Engobes for biscuit application are more like underfired glazes.

Usually engobes designed for plastic clay will not fit on bone dry or biscuit clay and vice versa.

Engobe composition

Engobe is made up of a mixture of plastic and nonplastic ingredients. Recipes for engobes look like those for glazes with a high percentage of refractory ingredients.

For white engobes, the plastic ingredients are china clay and ball clay. The amount of ball clay can be adjusted to get correct shrinkage.

The nonplastic ingredients are feldspar and quartz, and for low-temperature engobes frit is sometimes added to lower the vitrification point.

Small amounts of borax are often added to give better dry strength and to fuse the other ingredients together.

Engobes are often deflocculated like a casting slip, and in fact you can often use a casting slip which fires at the same temperature as your clay body.

Some typical engobe recipes are (from D. Rhodes: *Clay and Glazes for the Potter*):

TEMPERATURE RANGE:	Cone 08 – 1			Cone 1 – 6			Cone 6 – 11		
BODY CONDITION:	DAMP	DRY	BISC	DAMP	DRY	BISC	DAMP	DRY	BISC
Kaolin	25	15	5	25	15	5	25	15	5
Ball clay	25	15	15	25	15	15	25	15	15
Calcined kaolin		20	20		20	20		20	20
Leadless frit	15	15	15			5			5
Nepheline syenite				15	15	20			5
Feldspar							20	20	20
Talc	5	5	15	5	5	5			
Quartz (flint)	20	20	20	20	20	20	20	20	20
Opacifier (zircon)	5	5	5	5	5	5	5	5	5
Borax	5	5	5	5	5	5	5	5	5

10.3.3. COLOR OXIDE ADDITIONS TO ENGOBE

Coloring oxides are added to engobes as a percentage, as with glazes. However, since the color is diluted by the glaze over it, larger amounts are required. You should also remember that the oxide reactions will depend on the type of glaze being applied and on whether oxidation or reduction firing is used. Typical colors and oxide amounts are:

Red iron oxide

1- 5%	light green to light brown
5-10%	brown
10-15%	dark brown to black

Copper oxide or carbonate

1-5%	green or blue, red in reduction

Cobalt oxide or carbonate

1-5%	light to dark blue

Chrome oxide

1-5%	green

Manganese dioxide

1-10%	purple-brown

Nickel oxide

1-5%	gray or gray-green

Titanium dioxide or rutile

1-10%	tan, or mottled colors

Commercial glaze stains

1-50%	produces the color of the stain

As with glaze colorants, the most interesting colors are usually obtained by mixing combinations of oxides.

10.3.4. APPLICATION METHODS

A wide variety of decoration techniques can be used with engobe. Leather-hard ware permits the largest variety of techniques and usually has fewer technical problems compared to application on bone-dry or biscuit ware. Work flow for leather-hard engobe decoration is:

- Apply the engobe.
- Biscuit-fire.
- Apply the glaze.
- Glaze-fire.

Dipping, pouring

This is done the same as with glazes. The engobe should be just thick enough to completely cover the clay. Too thick application will often crack, especially on rims. If applied leather-hard, pouring and dipping should be done quickly, so that the pot does not get too soft from absorbing water.

Brushing

Brushing is one of the most satisfactory methods, especially for making bands or areas of engobe. The technique requires some skill in order to get an even coating. The brush should be fully loaded with engobe and should spread it evenly.

Spraying

The engobe must be thin enough to flow through the spray gun. It should be applied in several even coats, taking care to keep a smooth surface and to cover all areas equally.

Scratching or "sgraffito"

To get fine lines, engobe is applied to an area and, after it sets, it is scratched with a

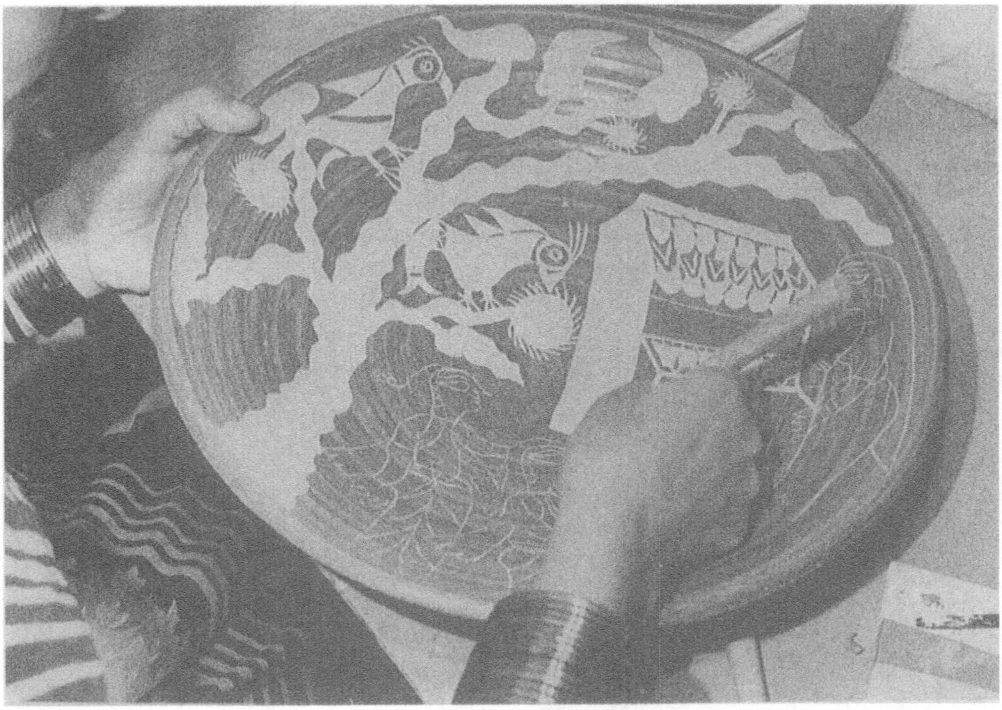

FIGURE 10.3.4.A A large bowl is decorated with a brush. Nepal.

sharp tool. This is called "sgraffito", which means "scratching". The clay color shows as a line.

FIGURE 10.3.4.B Sgraffito decoration. The white engobe is scratched off revealing the clay color.

Inlay

Lines are scratched on the leather-hard pot and then filled with engobe. The excess engobe is removed with a metal scraper after it sets, or with sandpaper after the pot is bone-dry. This results in a smooth surface, with the engobe lines contrasting with the clay color.

FIGURE 10.3.4.C White engobe inlay is scraped off after drying.

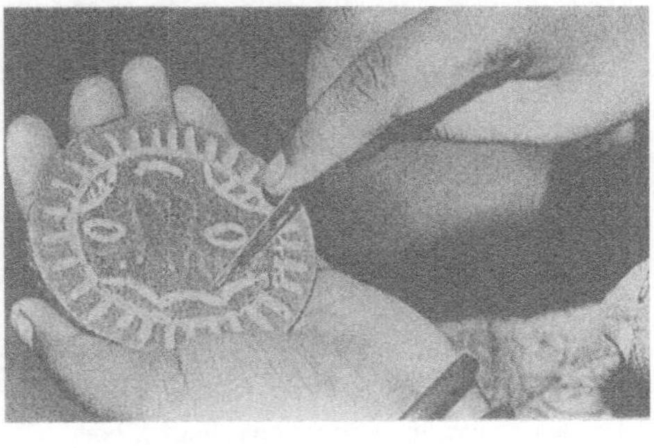

Stencil

Paper or plastic stencils are placed on the leather-hard pot, and engobe is brushed or sprayed over them. Afterwards the stencil is removed leaving the design of the stencil.

FIGURE 10.3.4.D Engobe stenciling.

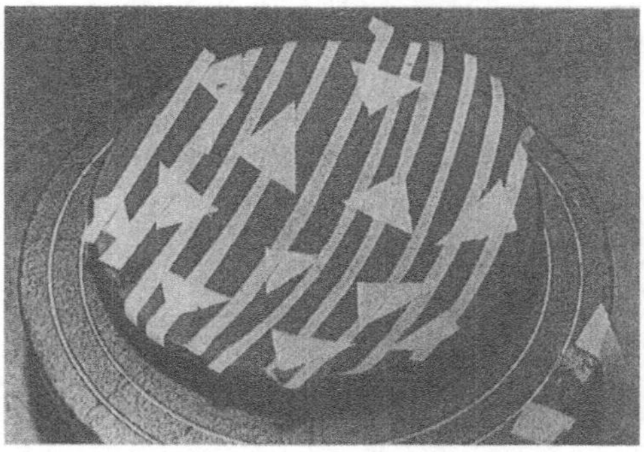

A) Paper is stuck on the plate in the desired pattern.

B) Engobe is brushed on.

C) The paper stencil is removed revealing the body.

Trailing

Usually called "slip" trailing, the engobe is applied by allowing it to flow from a device with a small opening, which produces raised line decoration. It is easiest to use a rubber bulb (such as an ear syringe available in pharmacies) or plastic containers used for soap or cosmetics. The opening can be made smaller by inserting small metal tubes.

FIGURE 10.3.4.E Slip trailing with an ear syringe.

Engobe hints

Engobes will show most clearly under a fully transparent glaze. However, semi-transparent or even opaque glazes can give beautiful effects, clouding the engobe colors.

Sophisticated decorators can take advantage of different glazes over the engobe to produce different colors. Complicated effects can result from applying different glazes to different areas of the decorated piece.

10.3.5. ENGOBE PROBLEMS

Often engobe will come off the pot. This almost always is caused by a different shrinkage rate of clay body and engobe and usually happens before firing. In many cases the engobe is applied too thick.

Engobe shrinks more than clay body

In this case, the engobe will develop cracks and will flake off, with the flakes curling away from the ware. The solution is to reduce the amount of plastic clay or substitute raw clay with calcined clay. Deflocculating usually helps.

Engobe shrinks less than clay body

In this case, the engobe will flake off, especially on rims and sharp edges, and the flakes will be flat. The solution is to add more plastic clay or to substitute calcined clay with raw clay.

Flaking after firing

This is caused by differences in firing shrinkage between clay and engobe. Usually adding flux to the engobe will help.

Spit-outs

Application of engobe to biscuit ware sometimes causes the engobe to lift off in small bubbles. This may only show up after glaze firing, but it arises during application. If the biscuit ware is very porous, it absorbs the water in the engobe so fast that air inside the body comes under pressure. When the air is released, it may blow out the engobe layer where the air escapes. The solution is to reduce the absorption by dipping the biscuit in water some time before engobe application.

10.4. TERRA SIGILLATA

The technique of coating pottery with terra sigillata was used by Roman and Greek potters and is still used by traditional potters in India and Nepal. It produces a thin, opaque and low gloss finish to pottery.

10.4.1. PREPARING TERRA SIGILLATA

Terra sigillata is made from clay. For temperatures below 1100°C local sedimentary clays are more suitable. The finer the clay particles the better. Such clays normally contain iron and fire to a red color. It is more difficult to produce white-firing terra sigillata from ball clay or kaolin.

by weight:	Clay	20
	Water	80
Sodium metaphosphate +		0.5%

The best result is obtained when ball milling the clay. Some clay can be prepared without ball milling. After ball milling the batch is transferred to a container and left for 24 hours. The coarse particles will settle and the upper 2/3 of the batch is siphoned off. A bucket with a tap placed 1/3 up is useful for regular production of terra sigillata.

Colors can be made by adding color oxides or pigments. First the terra sigillata is dried and the color oxide is added in amounts similar to what is mentioned for engobes (by dry weight). Then water is added and the batch is again ball-milled for 4 hours. It is then ready for use.

10.4.2. APPLICATION

The terra sigillata should be adjusted to a density of 1.15 to 1.20 for application on leather-hard clay. For dry and biscuit ware more water is added to obtain a density of 1.05 to 1.10. The ware should be clean and dust-free before application.

Application can be done by dipping, brushing and spraying. After drying the gloss can be improved by polishing the surface with a cloth.

10.4.3. ADVANTAGES

The use of terra sigillata makes it possible to produce attractive decorations on low-fired pottery without using glazes. The coating gives a dense, glossy and impervious surface. A very beautiful glossy black can be produced by placing the terra sigillata items in a lidded pot filled with sawdust. This is fired in a normal firing either in a kiln or in a traditional pottery firing. The strong reduction will change the normal red color to black.

The use of terra sigillata coatings as an intermediate layer between body and glaze is reported[2] to reduce crazing and bubbles in the glaze.

---

[2] W. G. Lawrence: *Ceramic Science for the Potter*, p. 214.

11. GLAZE PROBLEMS

Anybody with even a little experience with glazes will realize that problems often arise. Knowing what to do about them requires a lot of experience, and even expert glazers often find it difficult to establish the source of a problem.

11.1. INTRODUCTION TO GLAZE PROBLEMS

It is one thing to develop a nice glaze but quite another to keep it working. One potter may want a glaze that crazes, whereas another wants his glaze to be craze-free. A glaze fault may not mean that the glaze is ugly, just that it reacts differently and does not look like the desired effect.

When a glaze suddenly starts to react differently from what we want, we call it a glaze fault. Solving the problem is seldom easy, and usually several factors are involved. The first thing to check is what changes have occurred since the glaze last worked without problems. The following things should be checked:

1. Was the right recipe used and were the materials weighed out correctly? (Always use batch cards for glaze weighing).
2. Were the right raw materials used? Was there any chance of mistaking one material for another? Was the labeling of materials in stock correct?
3. Have new raw materials or a new frit batch been used since the last fault-free glaze batch was produced? If so, check if the new material is different from the original material in stock.
4. Has the body been changed in any way? For example: new preparation method, change of clay material, change of body recipe, higher or lower biscuit firing?
5. Was the preparation of the glaze done as usual? For example: same ball milling time, same screening, same specific gravity of glaze slip?
6. Was there any change in glaze application and were the products clean and dust-free before application?
7. Were there changes in the kiln setting? Was the glazed ware dry before firing started? Were there any changes in fuel, firing schedule, firing atmosphere (reducing/oxidizing) and was the correct top temperature reached (setting of cones, draw trials) ?

Once we know which conditions have changed, we may already be close to establishing what caused the glaze problem. The following trouble-shooting lists may be helpful to find solutions to the problem.

11.2. TROUBLE-SHOOTING CHECKLIST

11.2.1. GLAZE SLIP PROBLEMS

Problem: The glaze or part of the glaze settles too fast in the bucket.

Causes:
- High amount of frit in the glaze.
- Glaze materials too coarse.
- Water content of glaze slip too high.
- Ball milling time was too short.
- Too little clay in the slip.
- Metal buckets cause fast settling.

Solutions:
- Add 5-10% plastic clay or 0.5-2% bentonite.
- Reduce amount of frit by removing some of the insoluble materials from the frit recipe and adding these to the glaze batch instead.
- Longer ball milling of glaze materials.
- Add a small amount of vinegar (acetic acid) to the glaze.
- Use plastic buckets or wooden containers.

Problem: Glaze slip is too thin, low viscosity.

Causes:
- Water content too high.
- Alkali materials from frit or feldspar have been dissolved in the water, so the slip is deflocculated.

Solutions:
- Let the glaze stand for a day and decant the clear water off the top. Because some materials may be removed with the water, it is better to allow excess water to evaporate.
- Add flocculant (magnesium sulfate, calcium chloride), but only if the ware has too low a porosity. This is often done in production methods which use high-fired biscuit and lower temperature glaze.

11.2.2. PROBLEMS OF APPLICATION AND DRYING

Problem: Glaze layer too thin after dipping.

Causes:
- Water content of glaze slip too high.
- Clay body absorbs too little water.

Solutions:
- Increase density of slip (decrease water).
- Biscuit-fire at a lower temperature.
- Glaze only one side of the article at a time and allow it to dry before glazing the other side.
- Add flocculant to the slip so that glaze layer becomes thicker.

Problem: Glaze layer becomes too thick.

Causes:
- Glaze slip density is too high (too little water).
- The glaze does not contain enough clay materials.
- The biscuit body absorbs the water too fast.
- The glaze slip releases its water too fast.
- Dipping or pouring is done too slowly.

Solutions:
- Reduce density by adding water.

- Add plastic clay, bentonite or cellulose (CMC) binder.
- Biscuit-fire to a higher temperature or moisten the pots before glazing.
- Dip.

Problem: Glaze layer cracks during drying.

Causes:
- Glaze has too high a drying shrinkage due to high content of clay or zinc oxide or due to over-grinding in the ball mill.
- The glaze is applied too thickly.
- Single-fire glaze was applied to biscuit ware.
- In double glazing, the second glaze may tend to crack.
- Glaze is poured over a sprayed glaze.

Solutions:
- Less ball mill grinding of glaze materials.
- Calcine zinc oxide or add it to the frit.
- Replace part of the clay with calcined clay or introduce alumina (Al_2O_3) as feldspar.
- Reduce viscosity of glaze by adding water.
- Apply single-fire glaze to leather-hard pots.
- When double glazing, reduce clay content of the second glaze.
- When double glazing, apply the second glaze before the first one dries completely.

Problem: Glaze layer does not stick to the body.

Causes:
- Glaze adhesion to body is too low due to:
  greasy or dusty body surface
  too fast and too thick application
  too high body porosity
  too fine grinding of glaze
  dusty surface from underglaze colors.
- In single-fire glazing, the glaze shrinks less than the moist body.

Solutions:
- Clean the body surface by brushing or sponging.
- Reduce speed of dipping and, if spraying, apply less glaze at a time.
- Biscuit-fire at a higher temperature.
- Add cellulose binder (CMC) when single-fire glazing.
- Add clay, borax, soda ash or glaze to the underglaze colors.
- Reduce grinding time of glaze.

Problem: Glaze dusts off easily after drying.

Causes:
- Too little binding power of the glaze and low adhesion to body.

Solutions:
- Add 2-5% plastic clay or 0.5-1% bentonite.
- Add a binder like cellulose (CMC) or 1-2% borax.

11.2.3. PROBLEMS IN GLAZE MELTING

Problem: Glaze runs.

Causes:
- Firing temperature too high.

- Viscosity of glaze too low.
- Glaze layer too thick.

Solutions:
- Adjust firing temperature.
- Add alumina (clay, feldspar) or silica (quartz, zircon) to the glaze.
- Glaze thinner.

Problem: Glaze does not melt properly.

Causes:
- Firing temperature too low.
- Too much silica or alumina.
- Not enough glass formers (SiO_2 or B_2O_3) and to much CaO, MgO, BaO.
- Glaze materials too coarse.
- Evaporation of fluxes in the firing (extended firing time).

Solutions:
- Fire at higher temperature.
- Slower firing and longer soaking at the end.
- Increase amount of fluxes, reduce content of alumina.
- Longer milling of glaze materials.
- Add the evaporating fluxes to the frit.

Problem: Pinholes or eggshell surface.
This is one of the most common glaze problems and often the most difficult to cure. It is usually caused by gas escaping from the body or glaze, leaving small holes that do not have time to smooth over.

Causes:
- Firing temperature too low, glaze does not have time to melt completely.
- Firing temperature too high, glaze reacts with body, forming gas bubbles.
- Glaze has high surface tension and viscosity, which do not allow gas bubbles to escape.
- Glaze is too thick, not allowing gases to escape.
- Release of gases from body or engobe.
- Early reduction firing forms carbon and sulfates in the body, which cause pinholes when they are later released.
- Body contains organic particles which burn out, leaving small pits.
- Body contains air bubbles from incorrect slip casting.
- Glaze contains zirconium opacifier, which causes large amounts of gas.
- Contamination from ball milling, usually by linings set in lime cement.
- Contaminated water used to mix glazes.
- Dirty biscuit or dust on the glaze.

Solutions:
- Fire to the correct temperature.
- After final glaze temperature is reached, "soak" the kiln by holding at the same temperature for about half an hour.
- Reduce viscosity and surface tension by changing the glaze recipe.
- Less reduction, especially in early stages of firing.
- Thinner glaze application.
- Higher biscuit firing.
- Body must be prepared to eliminate organic materials (sometimes long aging to decompose the material and repugging will solve the problem).
- Slip must be mixed and poured carefully to eliminate air bubbles.
- Increasing viscosity of zirconium-opacified glazes by addition of clay or talc may reduce the problem.

- Ball mill linings should be fastened with high-alumina cement mortar.
- Use only clean water to mix glazes.
- Biscuit should not be stored too long and should always be cleaned before glazing. Glazed ware should be put in the kiln as soon as possible.

Problem: Glaze crawls.

Normally this is caused by cracking or lifting of the glaze layer before firing (see section 11.2.2. for additional causes and solutions).

Causes:
- The viscosity and surface tension of the melted glaze are too high.
- Too fast firing.
- During firing, the body released gases (steam, carbon, sulfur), which lifted the glaze layer off.
- Drying and sintering shrinkage high.
- Glazed ware was still wet when fired.

Solutions:
- Reduce surface tension and viscosity by firing higher.
- Reduce alumina, magnesia and zircon.
- Increase biscuit temperature.
- Correct what causes cracking and lifting of dry glaze (see above).
- Reduce grinding time of glaze.
- Reduce drying and sintering shrinkage by calcining part of clay and zinc oxide content.
- Dry the glazed ware before firing.
- Add 1-2% borax to the glaze.
- Add clay, borax or frit to underglaze colors.

Problem: Glossy glaze turns matt.

Causes:
- Glaze is absorbed by the body (glaze layer is too thin).
- Flux materials evaporate in firing.
- Sulfates from fuel are deposited on the glaze surface.
- Too much steam in the kiln.
- Glaze is underfired.
- Crystals form due to very slow cooling.
- Glaze slip was not properly mixed.

Solutions:
- Glaze thicker.
- Glazed ware set next to biscuit ware or on new shelves, which may attract volatile fluxes from the glaze. Do not mix glaze and biscuit firing.
- Introduce volatile fluxes in the frit.
- Fire with more draft and less reduction and cool quickly.
- Always stir glaze immediately before application and screen it through at least 60 mesh.

Problem: Matt glaze turns glossy.

Causes:
- Glaze is overfired.
- Too fast cooling.
- Oxides from the body combine with the glaze.

Solutions:
- Slower firing, or longer soaking period at the end of the firing.
- Cool slowly, closing all dampers and fireboxes.
- Use another matting agent.

Problem: Glaze color changes.

Causes:

- Wrong firing temperature, often overfiring.
- Change in reduction/oxidation atmosphere.
- Impurities in glaze or body.
- Coloring oxides not ground fine enough.
- Color pigment from engobe or underglaze melts into the glaze.

Solutions:

- Better control of firing temperature, oxidation/reduction.
- Check materials for impurities.
- Control the grinding time of coloring oxides.
- Add clay or quartz to underglaze pigments and engobes.

11.2.4. PROBLEMS AFTER FIRING

These are problems that only appear right after firing or after the article has been used for some time.

Problem: Crazing of the glaze

Causes:

- Thermal expansion of the glaze is higher than the body, which causes the glaze to contract more in cooling. This may be caused by:
- Too much alkali (soda and potash).
- Too little silica, alumina or zinc oxide.
- Too little cristobalite formation in the body.
- Lack of strong clay/glaze interface.
- Glaze is too thick.
- Fast cooling.

Solutions:

- Apply glaze more thinly.
- Higher glaze firing temperature (to increase cristobalite).
- Higher biscuit firing temperature.
- Reduce amount of soda and potash in glaze by replacing with boron (to decrease thermal expansion of glaze).
- Add additional boron, silica, zinc oxide or calcium carbonate to the glaze (to decrease thermal expansion of glaze).
- Add quartz or talc to the body. Quartz forms cristobalite in the body, which has a high thermal expansion.
- Longer soaking at top temperature, slower cooling.

Causes:

- Moisture expansion (sometimes called delayed crazing). Porous bodies expand when they absorb moisture from the air and force the glaze to craze. This is a common problem with earthenware.

Solutions:

- Fire at higher temperature or add flux to the body, making the body more vitreous.
- Add calcium carbonate, talc or dolomite to the body.
- Reduce the thermal expansion of the glaze.

Problem: Shivering of glaze

This is usually seen as particles of glaze falling off the pot after firing (sometimes after a few weeks). It happens most often on sharp edges, but the entire glaze may shiver or the pot may crack.

Causes:

- Thermal expansion of the glaze is less than the body, which leaves the glaze under strong compression. This may be caused by:
- Too high content of silica, boron or zinc oxide in the glaze.
- Too high content of silica in the body.
- Too low content of soda or potash in the glaze.

Solutions:

- Reduce the amount of quartz in the glaze.
- Reduce the amount of quartz in the body.
- Add more soda or potash to the glaze.

Problem: Lime popping

This is seen as small pieces of glaze popping off the pot, often several weeks or even months after firing. Under each flake of glaze a small white particle can be seen imbedded in the body.

Causes:

- Small pieces of limestone or plaster in the clay body. These slowly absorb moisture from the air and expand, forcing the glaze off the pot.

Solutions:

- Find the source of the lime pieces.
- Replace contaminated material or screen it through 40 mesh.
- Plaster that has gotten into the clay. All contaminated clay must be thrown out, and better care taken in clay mixing. The problem often comes from recycled clay, which has picked up plaster in the forming section.

11.3. SPECIFIC PROBLEM EXPLANATIONS

11.3.1. CRAZING AND SHIVERING

As already mentioned, crazing and shivering are caused by differences in thermal expansion/contraction between the glaze and body.

Crazing

Crazing appears as cracks in the glaze. This occurs during cooling, if the glaze contracts more than the clay. Cures for crazing are mentioned above.

Shivering

Shivering is the opposite of crazing and occurs during cooling if the clay contracts more than the glaze. Cures for crazing are mentioned above.

Thermal expansion

All materials expand when heated. This is called thermal expansion. Some materials expand more than others, and the degree of expansion can be measured. Numbers are used as a scale of thermal expansion, and this is called the *coefficient of expansion* (CE). The glaze layer on a pot has one coefficient of expansion and the body has another.

Glaze-body tensions

After a pot is fired and taken out of the kiln, it will be exposed to a sudden decrease in temperature. The glaze layer and the body will contract, but most often at different rates. Below is shown what happens when 1) glaze contracts more than body and 2) body contracts more than glaze.

FIGURE 11.3.1.A Thermal expansion tester. The small electric kiln (1) is shuttled to the left while the test material is placed in a holder (2) next to a thermocouple. Expansion is read on a micrometer (3) and temperature on the voltmeter (4).

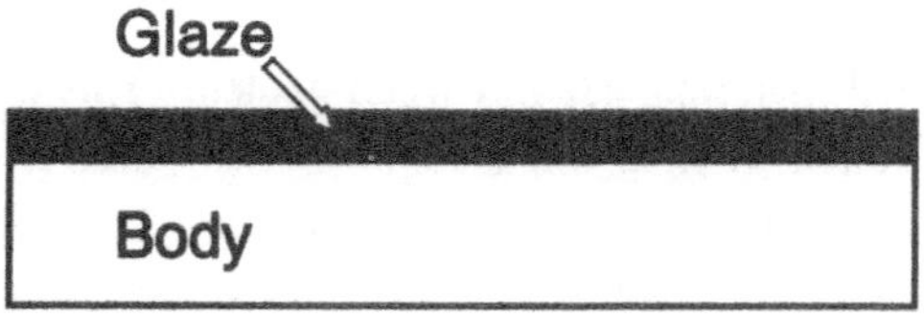

This figure shows a body (white) with a glaze on top (black). The glaze and the body have contracted at the same rate. We say: they have the same coefficient of expansion (CE).

Glaze contracts more

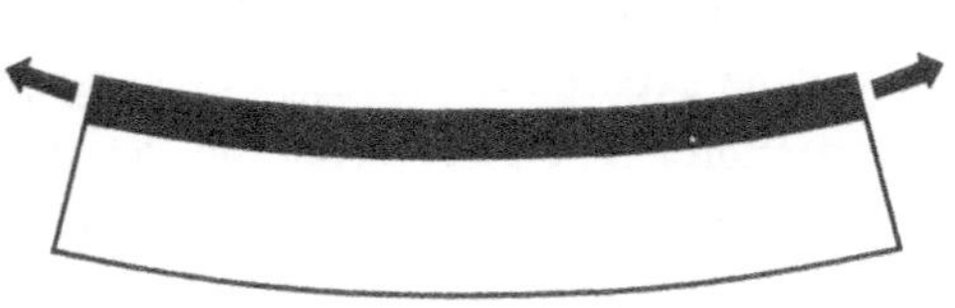

This figure shows a glaze that has a higher coefficient of expansion (CE) than the body.

The glaze contracted more, so it is shorter and therefore the glaze is under a tensile stress (it is pulled apart). If the body is very thin it will bend as shown. The arrows show the direction of the stress the glaze is under.

More often the tensile stress is relieved by cracks in the glaze as shown in this figure. This is called crazing. The stress caused by high CE of the glaze may be relieved by crazing as soon as the pot is taken out of the kiln or it may take days, months or years. The longer it takes, the closer is the CE of body and glaze.

Body contracts more

This figure shows a body with higher CE than the glaze. The body contracted more than the glaze. The glaze is under compression, and if the clay is thin it may bend as shown to relieve the pressure. If body contraction is only slightly greater than glaze contraction, nothing will happen.

If a glaze contracts much less than the body, the compression on the glaze becomes too much and the glaze will start to flake off like this (shivering). This may not happen by itself, but only if something hits the pot. Typically, the rim of a cup will easily chip off.

High compression of the glaze may also be relieved by cracking of the body.

Moisture swelling

When the body has been exposed to humidity for a long period, water enters the body, which expands slightly (moisture swelling). This expansion causes the glaze to go into tension and it will craze. This kind of crazing is called delayed crazing.

Solutions

As we saw above, crazing and shivering are caused by different rates of contraction and expansion (different CE's). The problems are cured by adjusting the CE of body and glaze, so that the two contract and expand more closely. It is best if the glaze is left under slight compression.

Coefficients of expansion for various materials

Ceramic materials have different coefficients of expansion (CE). The following list shows the relative values for the most common:

Na_2O	High CE	
K_2O	↑	contracts *more*
CaO		in cooling.
BaO		
TiO_2		
Fe_2O_3		
Al_2O_3		
PbO		
CuO		
MnO		
ZrO_2		
SnO_2		
P_2O_5		
ZnO		
MgO		
SiO_2	↓	contracts *less*
B_2O_3	Low CE	in cooling.

Adjusting CE of glaze

From this list we can see that if we replace soda (Na_2O) with boron (B_2O_3) in a glaze we will lower the CE of the whole glaze. This can be done without changing the melting point of the glaze. Addition of silica will lower the glaze's CE but will also raise its melting point.

If shivering occurs, it means the CE of the glaze is too low. Adjusting it means adding soda (Na_2O) and reducing boron (B_2O_3).

Adjusting body

Adjustment of body CE is not done according to the CE of the materials listed above. The contraction rate of the body depends to a much higher degree on the sudden reversible contraction of silica crystals when these change their crystal structure (cristobalite).

Quartz change

Quartz is a crystal form of silica. Quartz is created in the body during firing when the clay crystal changes form and releases some of its silica. When quartz is heated it changes its crystal structure at 573°C. This happens very suddenly and is accompanied by a 1% expansion. On cooling to below 573°C it contracts again.

Cristobalite

Cristobalite is another crystal form of silica. It changes its size around 220°C and the volume change is nearly 3%. Cristobalite is created at temperatures above 900°C from silica released from the clay ($Al_2O_3 \cdot 2SiO_2$) or talc ($3MgO \cdot 4SiO_2$) or from quartz.

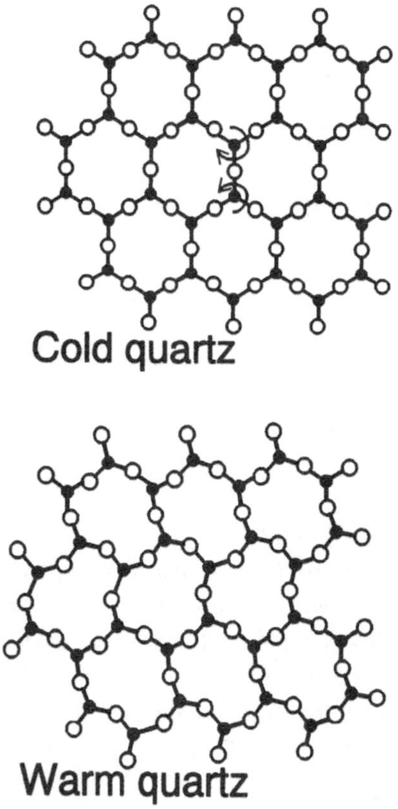

FIGURE 11.3.1.B The volume change of quartz is caused by a rearrangement of the bond between the atoms. At 573°C the angle suddenly shifts as shown.

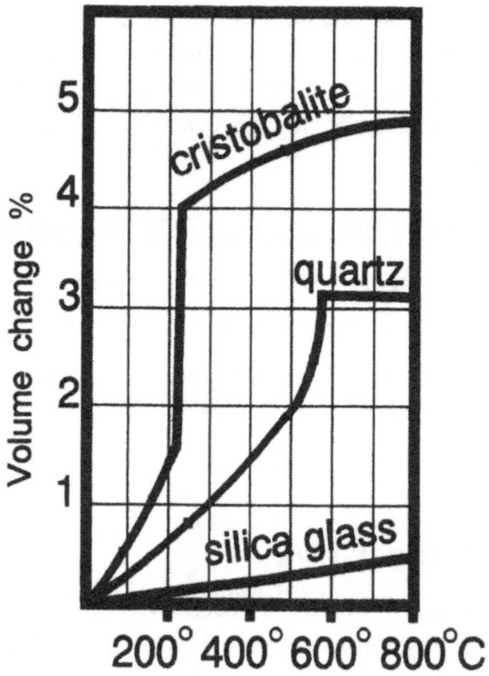

FIGURE 11.3.1.C The graph shows volume changes of three forms of silica. The two crystal forms change dramatically but silica in glass hardly changes.

Body-glaze contraction

The two graphs below show how the body and its glaze contract during cooling. The graph in Figure 11.3.1.D shows a body that does not contain any cristobalite. At 573°C the body contracts suddenly due to the contraction of quartz, but at this temperature the glaze is still fluid enough to follow the contraction of the body.

FIGURE 11.3.1.D Contraction of body and glaze during cooling. Glaze contracts more so it will craze.

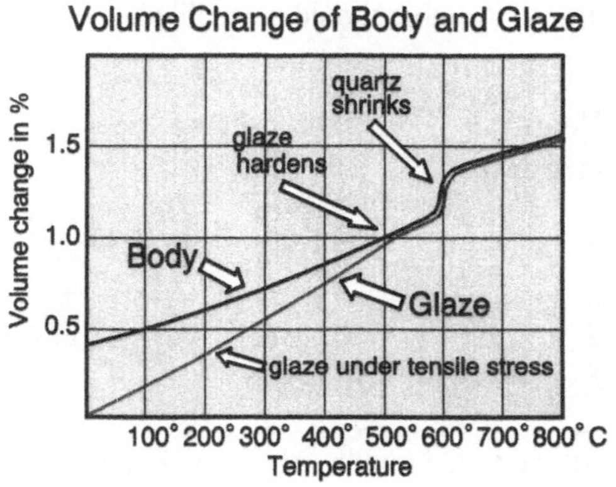

Around 500°C the earthenware glaze hardens and from then onwards contracts according to its own CE. In this example the glaze has a higher CE than the body; it contracts more. This leaves the glaze under tensile stress; the glaze is smaller than the body. This will cause the glaze to craze.

FIGURE 11.3.1.E The body in this graph contains cristobalite and shows a sudden contraction at 220°C. This causes an overall higher contraction of body compared to glaze.

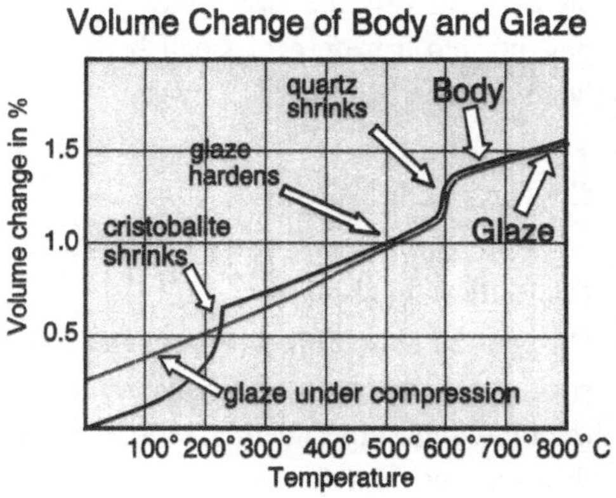

The graph in Figure 11.3.1.E shows contraction of a body containing cristobalite. As above, the glaze first follows the quartz contraction, then hardens and starts to contract more than the body. However, at 220°C the cristobalite change causes the body to contract, and at this temperature the glaze is hard so it is left under compression. This compression will prevent the glaze from crazing.

Moisture crazing

After firing, the porous earthenware body will absorb moisture and this causes the body to expand. If the glaze is not under sufficient compression it will craze. Such delayed crazing may occur a long time after firing. The moisture expansion of the body is reduced by making the body more vitreous. Additions of talc or limestone to the body reduce moisture crazing.

Crazing cure

For both types of crazing the cure is:

- Add quartz (or silica), talc or limestone to the body.
- Biscuit-fire to a higher temperature.
- Glaze-fire to a higher temperature.
- Add silica to the glaze.
- In the glaze, replace fluxes with high thermal expansion, like soda (Na_2O) and potash (K_2O), with boron oxide (B_2O_3).

It may seem strange that the cure for crazing is to add silica to both body and glaze. The reason is that adding silica to glaze makes it contract less, but silica added to the body causes the body to contract more.

Crazing test

There are several ways to test how the expansion of glaze and body fit each other. The most simple ones are:

- Rings of clay with a diameter of 5 to 10 cm are made with a small gap and biscuit-fired. The gap is measured, the ring is glazed on its outer surface and refired. After firing the gap is measured to see if the ring has contracted or expanded. If the gap has become greater the glaze will craze.

- Glazed samples are exposed to thermal shocks by repeated heating and cooling.

The thermal shocks can be from boiling water into ice water. The number of cycles the sample can withstand before crazing indicates its craze resistance.

Another method is to heat the sample at first to 100°C then cool it in 20°C water. This is repeated while raising the temperature in steps of 10 or 20 degrees. The higher the heating temperature the sample withstands without crazing the longer it will be able to stay craze-free under normal conditions.

A rough guide is

120°C	craze-free for	8 days
150°C	craze-free for	100 days
180°C	craze-free for	2 years
200°C	craze-free for	life

Even if a sample survives the thermal shock test it may still craze due to moisture swelling. This can be tested in an autoclave which is simply a pressure cooker that can withstand higher pressures. A pressure cooker can be used instead. The glaze sample is placed in the pressure cooker with some water. It is kept under pressure for a period and then checked for crazing. The time it can withstand pressure without crazing indicates the time it may stay craze-free under normal circumstances. The following rough guide is for testing in an autoclave under a pressure of 3 atmospheres (about 3 bars). If using a pressure with, say, a pressure of 1.5 atmospheres the testing time in the table should be doubled:

Hours in autoclave	Expected craze-free life
1	1- 2 years
2	2- 3 years
3	4- 6 years
4	9-10 years
5	13-15 years

All the tests provide only a rough indication of craze resistance. When you do the test you will develop your own procedure, which then should always be followed faithfully. In this way you will be able to compare your crazing test with your previous results.

11.3.2. CRAWLING

Crawling appears as areas of clay that are not covered by the glaze. It may be small areas or, in extreme cases, the glaze may pull up into a pattern of small balls or islands, leaving bare clay in between.

Crawling is caused by:

Poor adhesion of glaze

Dusty or oily biscuit prevents the glaze from sticking to the body. Refractory oxides (chrome, rutile) or underglazes that act as a dust layer prevent the formation of an inter-

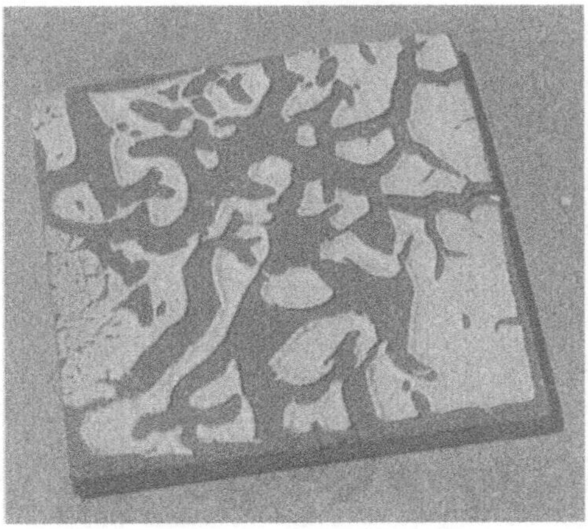

FIGURE 11.3.2.A Excessive crawling on a tile.

face. Adding clay, borax or frit to the underglaze colorants helps.

High surface tension

High surface tension of the glaze in melting pulls it into islands before the clay/glaze interface forms. This is caused by certain oxides, especially magnesia, clay and zinc oxide. The solution is to replace magnesia by other materials, to calcine part of the clay or to use calcined zinc oxide instead of raw.

Cracking of glaze layer

Extensive shrinkage of glaze in drying and early stages of firing, usually caused by too much clay content or by overgrinding the glaze, causes the glaze to crack and separate from the body. A thick glaze layer is more likely to crack.

11.3.3. PINHOLING AND BLISTERING

Pinholes appear as tiny holes in the glaze surface. Blisters look like frozen bubbles or craters. They are a problem in utilitarian ware, as they collect dirt. They may be only

FIGURE 11.3.3.A Pinholing.

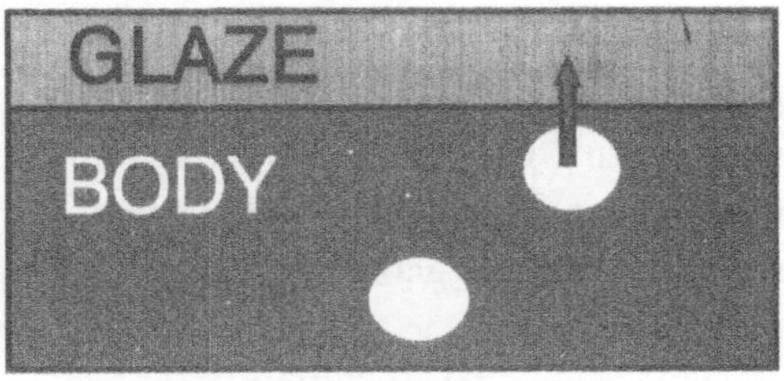

a) Section of glaze and body with two air holes. The one close to the surface rises.

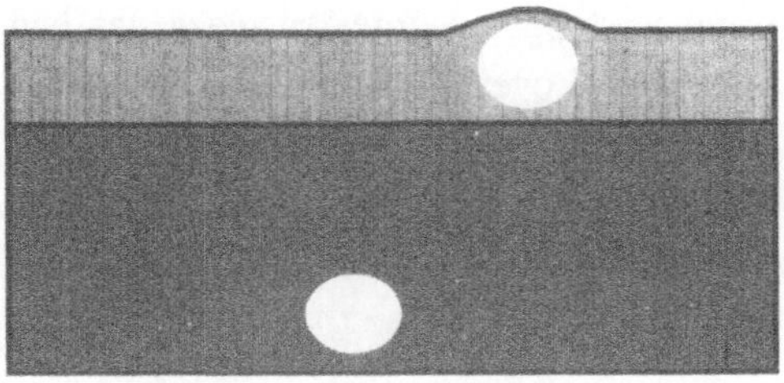

b) The air bubble pushes through the melted glaze.

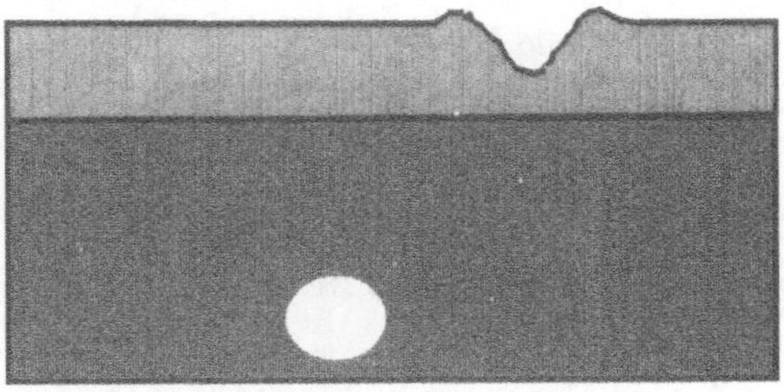

c) The bubble has burst and left a crater.

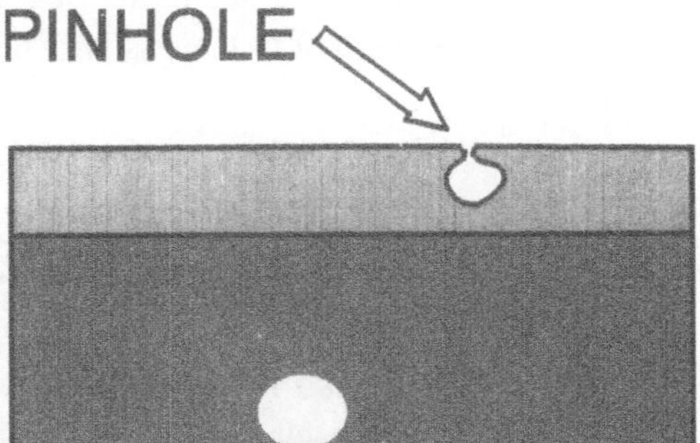

d) The melted glaze has partly moved back over the crater, but firing was stopped too early and a pinhole remains.

on the surface of the glaze or may penetrate to the clay layer.

Pinholes

During firing gas bubbles are formed in the melted glaze. The bubbles will move to the surface of the fluid glaze and be released.

If you watch any glaze metling, you can actually see this process. Some glazes (especially those containing raw borax) foam and boil until they finally smooth out. When the firing is stopped before the glaze has had time to heal over, a pinhole or crater is left (see Fig. 11.3.3.A). Since overfiring also causes pinholes it is better to keep the maximum temperature for some time (soaking period).

The main sources of the gas are:

- After glazing, a large volume of air exists in the space between the solid glaze materials. The air gathers into bubbles during sintering and melting.
- Release of sulfates and carbon in the body and from some of the glaze materials.
- Air bubbles in the body introduced by improper handling of the casting slip.
- Sulfates and carbon from the fuel may deposit in the body during the initial stages of firing. Above 900°C the gas will be released.

It is important to find out if the problem is in the glaze or in the body. Relatively large pinholes that go all the way to the body are usually caused by small holes in the body that do not accept the glaze – this is most common with slip-cast ware, or with common red clay that contains particles of organic matter, sand or mica.

Problems arise if the glaze starts to cool and solidify while bubbles or craters are still forming.

Detailed causes and solutions are given in section 11.2.3.

11.3.4. COLOR CHANGES

Potters are often plagued by changes in glaze color, either within the same kiln-load or from separate firings.

Often this problem can be traced to glaze preparation. The colors may not be ground finely enough, weighing may be incorrect, raw materials may have changed.

Otherwise, the problem usually is due to more or less reduction than usual. This is one of the most difficult conditions to control in firing and depends completely on the skill of the firemaster.

The problem is worst in glazes that contain color oxides that are sensitive to reduction. The most sensitive is copper, which is green in oxidation and red in reduction; and iron oxide, which is yellow, red to brown in oxidation and mottled red-brown, gray to blue or green in reduction. Other oxides change less.

Heavy reduction will darken the iron in the body, which will affect the glaze, also darkening it. Sometimes a pot will be dark on the reduced side and light on the oxidized side.

Other causes and solutions are given above.

12. DEVELOPING GLAZES

Most potters do not know much about glaze chemistry and are usually afraid to develop their own glazes. It is true that glaze chemistry is difficult to understand without a background in chemistry. Still, a simple knowledge of fluxes, stabilizers and glass formers and how they combine to make glaze is useful. It will give even the non-technical potter an idea of how to approach glaze problems and develop new colors and textures. The reader with more technical background can make use of the Seger formula for a more sophisticated approach to glazes.

FIGURE 12.0.0.A Preparing glaze tests in the ceramic laboratory.

Remember that glaze chemistry is something new in the history of ceramics. Before it existed, potters found out how to make glazes by trial and error, without modern methods of analysis or even accurate methods of weighing.

It is also true that once glaze recipes were developed they were closely guarded secrets. A special glaze that no one else could duplicate gave an advantage over the competition. The modern systematic, scientific approach to glaze development has largely made "secret" glazes obsolete because the action of the various glaze ingredients is now well known.

On the other hand, glaze making is very much like cooking. The same recipe may produce very different results when prepared by two different cooks! Standard glaze recipes as published in books often do not work because of differences in raw materials, firing technique etc. Just as a good cook should know how to substitute materials, a good glaze maker can develop an intuitive knowledge of glazes. Like most things, this only comes from hard-won experience.

Potters should not be afraid to experiment with glaze development, although most of them simply do not have the time to take away from production. The following chapters describe standard approaches to glaze development and will be useful to those who want to experiment with glazes.

12.1. MODIFYING EXISTING GLAZES

The easiest approach to working with glazes is to start by modifying existing glazes. These may be new recipes from books, or recipes that you are already using. Although you can take a trial-and-error approach, this takes a lot of time and money, and results are largely a matter of luck. It is best to use some of the systematic methods presented below and knowing even a little about the nature of the various glaze materials will help you reach your goal more efficiently.

There are three approaches to modifying glazes:

– You have a goal for a particular surface, color, texture etc.

– You have a glaze problem that needs to be corrected.

– You have no special goal but just want to see what happens when you change the recipe.

12.2. BASIC EQUIPMENT

The minimum equipment required for glaze testing is:

FIGURE 12.2.0.A Triple beam balance for weighing up to 2.6 kg.

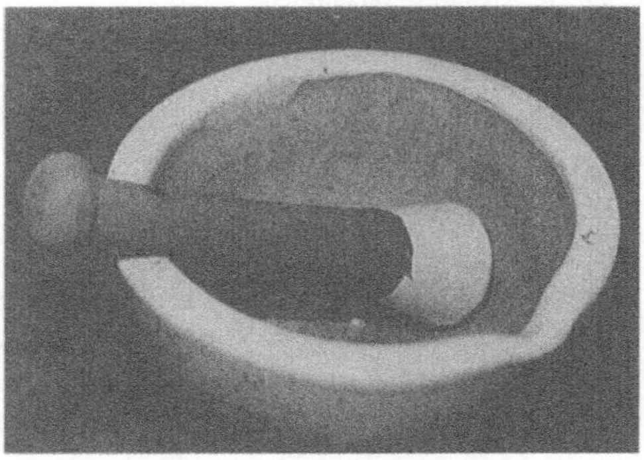

FIGURE 12.2.0.B Mortar and pestle for grinding materials.

- Accurate balance, either a triple beam balance or a goldsmith's balance with weight. Spring-type scales are not accurate enough for weighing glazes.
- A mortar and pestle for grinding materials. These should be porcelain, so that they do not contaminate tests.
- A 100-mesh sieve. This can be bought ready-made, or you can make your own using brass or stainless steel screen.
- Firing can be done in your regular glaze firing.

12.3. TESTING METHODS

12.3.1. TEST PIECES

The type of test piece you use is largely a matter of personal preference. In order to show glaze behavior under various circumstances, it should have:

- A horizontal and vertical surface.
- A textured area.
- A hole to tie it. This helps to keep similar tests together for future reference.
- An area for labeling. It is best to write the full recipe of the test on each test tile (with a brush and iron oxide and water, or chrome oxide or engobe etc.) since code numbers often get confused and notebooks get lost.

The first step in understanding your materials is to fire all available materials in small bowls at your standard glaze firing temperature.

A small quantity of each material (ground and sieved through 100 mesh) is placed dry in the bowl. This will show you which ones melt alone, and which ones remain as powder. Most of the materials will not melt but may change in color or may react with the clay. Only the strong fluxes will melt by themselves – other materials that do not melt may be fluxes, but only in combination with other materials.

12.3.2. LINE BLENDING

Line blending is a systematic way of finding out the reactions of two different materials (or mixtures of materials).

The easiest way is to prepare the two materials by grinding, sieving and mixing them with water in two separate containers. To make the line blend, the materials are mixed by volume (using a small spoon) and applied on a test tile, starting with one material alone and adding the other material in equal steps. Since the tests are measured by volume it is important that the same amount of water is added to the two line blend materials.

Glaze half of the test tile twice to show variation in glaze thickness.

MATERIAL	PARTS BY VOLUME (number of spoonfuls)										
material A	0	1	2	3	4	5	6	7	8	9	10
material B	10	9	8	7	6	5	4	3	2	1	0

Line blend 10 steps

An example of a line blend in 10 steps, which gives the full range of combinations of 2 materials, is shown in the table above.

The most common use of the line blend is to find out the effect of one material in a standard glaze recipe. If material A is the standard glaze recipe, material B could be the standard glaze + a coloring oxide addition of 5-10%.

Line blend 5 steps

Usually, 5 steps will be enough for the first test. In this example material A is the basic glaze, material B is the basic glaze + 10% copper oxide.

	PARTS BY VOLUME					
TEST NUMBER	A	B	C	D	E	F
GLAZE	0	1	2	3	4	5
GLAZE + 10% CuO	5	4	3	2	1	0
CuO in test	10%	8%	6%	4%	2%	0%

Mixing procedure

- First prepare a line blend mixing card like the one above.
- Prepare the two mixtures as usual and add the same amount of water to each.
- Place the two materials in bowls in front of you, glaze A to your left and glaze B to your right.
- In the middle place an empty bowl into which you pour the spoonfuls from the two other bowls according to the number for each test on your line blend card. Keep track of the spoon counting by marking the line blend card.
- Stir the test mixture well.
- Mark a test tile with the mixture's test number (date + serial number).
- Glaze the test tile.
- Discard the remaining glaze and continue with the other line blend mixtures.

Calculation example

In the case above it was easy to calculate the copper oxide addition in each of the tests. When more complex mixtures are used in a line blend the calculation becomes more complicated. Here is an example of mixing two glazes:

Glaze A:	Frit X	70
	feldspar	15
	quartz	5
	kaolin	10
Glaze B:	Frit Y	80
	zircon	10
	kaolin	10

	PARTS BY VOLUME					
TEST NUMBER	A	B	C	D	E	F
GLAZE A	0	1	2	3	4	5
GLAZE B	5	4	3	2	1	0

After firing, test number D turned out to be the most interesting. We now want to test a larger amount of this and the recipe is calculated in this way (see table next page):

TEST D	PARTS	FRIT X	FRIT Y	FELDSPAR	QUARTZ	ZIRCON	KAOLIN
GLAZE A	3	210	–	45	15	–	30
GLAZE B	2	–	160	–	–	20	20
TOTAL PARTS	5	210	160	45	15	–	50
NEW GLAZE D	1/5	42	32	9	3	4	10

Materials in glaze A were multiplied by 3 and those in glaze B by 2.

The sums of each material were then divided by 5 and the final recipe is:

Test D:		
	Frit X	42%
	Frit Y	32
	Feldspar	9
	Quartz	3
	Zircon	4
	Kaolin	10

The recipe is based on a line blend test measured by spoonfuls. That is not very accurate so, before going any further, the test result should be retested by weighing the dry materials.

12.3.3. TRIAXIAL BLENDING

Triaxial blending is a method of testing varying amounts of three different materials or colors.

FIGURE 12.3.3.A Display of results of triaxial testing with overglaze pigments.

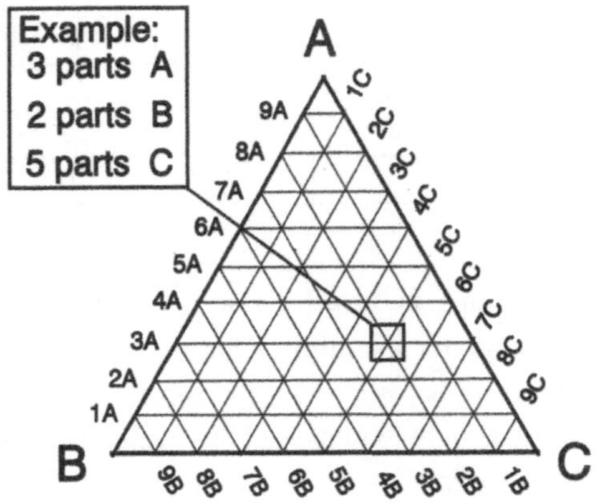

FIGURE 12.3.3.B A triaxial blending chart system with 10 steps. Composition of a test at an intersection is found by following the lines to the periphery of the triangle.

Each corner of the triangle represents 100% of the material. Each side of the triangle is the line blend of the materials at its ends, and the intersections inside the triangle represent combinations of all three materials. So the result is three line blends, plus all the combinations. Fig. 12.3.3.B is an example of a triaxial system with 66 tests.

The system is better explained by an example. You may have a basic opaque

FIGURE 12.3.3.C Triaxial system with 5 steps. The number at each point refers to the test number on the triaxial blending card.

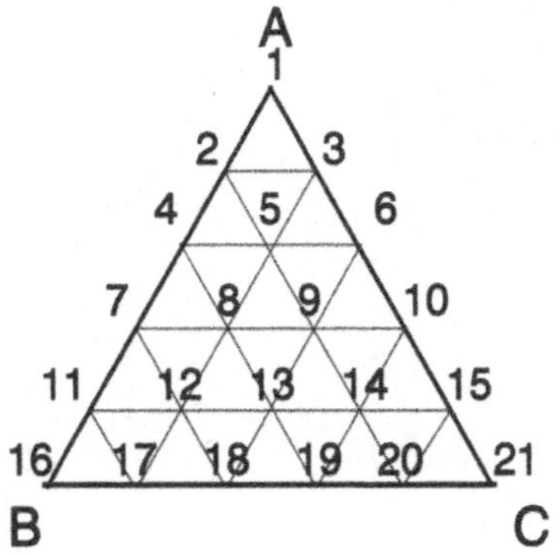

glaze and you want to see how it responds to 3 different coloring oxides: cobalt oxide, copper oxide and iron oxide. In this case we use a simple triaxial blend with only 21 tests as shown in Fig. 12.3.3.C.

The procedure is:

- Prepare a triaxial blending card as shown.
- Prepare 3 mixtures of basic glaze with oxide additions:

A	glaze + 5%	cobalt oxide
B	glaze + 10%	iron oxide
C	glaze + 10%	copper oxide

- Add same amount of water, screen 100 mesh.
- Place 3 bowls with the mixtures in front of you: B on the left, C on the right, A in the center. Right in front of you place an empty bowl.
- Have all test tiles numbered and arranged in sequence near by.
- Collect teaspoonfuls of each mixture; A,B,C according to the numbers on the triaxial card. Mark each time you have finished collecting from each bowl.
- The mixture is collected in the empty bowl.
- Stir the mixture, pick the test tile with the right triaxial blend number.
- Glaze the test tile.

FIGURE 12.3.3.D Arrangement of bowls for triaxial mixing.

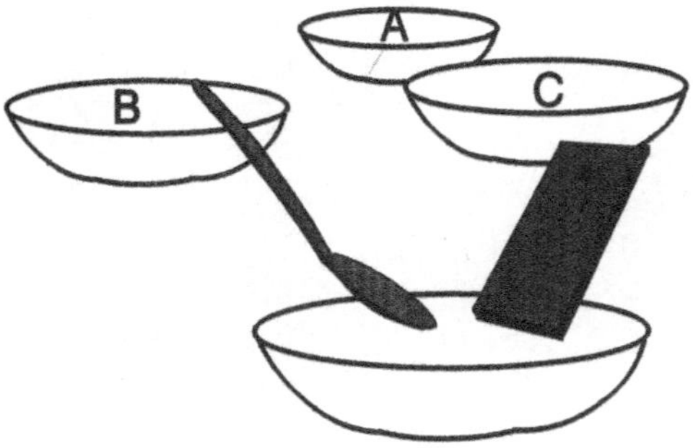

Getting the right number of spoonfuls into the collection bowl for each test takes a lot of concentration. A mixing card as shown below helps you to keep track of your progress with the spoon counting.

TRIAXIAL BLENDING CARD																					
TEST NUMBER	1	2	3	4	5	6	7	8	9	10	11	12	13	14	15	16	17	18	19	20	21
MATERIAL	NUMBER OF SPOONFULS																				
MIXTURE A	5	4	4	3	3	3	2	2	2	2	1	1	1	1	1	0	0	0	0	0	0
MIXTURE B	0	1	0	2	1	0	3	2	1	0	4	3	2	1	0	5	4	3	2	1	0
MIXTURE C	0	0	1	0	1	2	0	1	2	3	0	1	2	3	4	0	1	2	3	4	5

If you want to know the recipe of one of the tests, say number 14, you calculate this in the same way as for line blends:

	PARTS	GLAZE	COBALT O.	IRON O.	COPPER O.
MIXTURE A	1	100	5	–	–
MIXTURE B	1	100	–	10	–
MIXTURE C	3	300	–	–	30
TOTAL	5	500	5	10	30
D RECIPE	total/5	100%	+ 1%	+ 2%	+ 6%

Once you get used to working with triaxial blends, you will be able to read the percentage directly from the triangular chart.

This triaxial blend was based on only 21 variations. Out of these only 6 were blendings of all three mixtures; the rest were simply line blends involving only two mixtures. A larger triaxial blend system would produce more intermixing of all three materials, but also a lot of extra work. However, you could use a system with 10 steps on each side as shown in Fig. 12.3.3.B, but leaving out the line blends A-B, A-C and B-C, and only blend the 36 tests in the center of the triangle.

12.3.4. KEEPING RECORDS

The key to experimenting with glazes is keeping accurate records and labeling them in such a way that the actual tests can be compared with your notebook.

As mentioned above, it is best to write the entire recipe on the test tile itself, along with the date of testing. This is possible with a simple test like adding coloring oxides to your basic glaze. For more complicated tests and line blend or triaxial blend tests, you will have to rely on the test number on the tile. Mark the date, the number and, if you do more than one test a day, add a serial number. In your notebook, the date and recipe are also written.

Make it a habit to take notes of the fired results immediately after unloading the kiln. Write down firing conditions, location of test tile in the kiln, and your impression of the glaze. Is it well melted, running, pinholes or tendency to crawl? Use a whole sheet of paper for each test or test series. Fi-

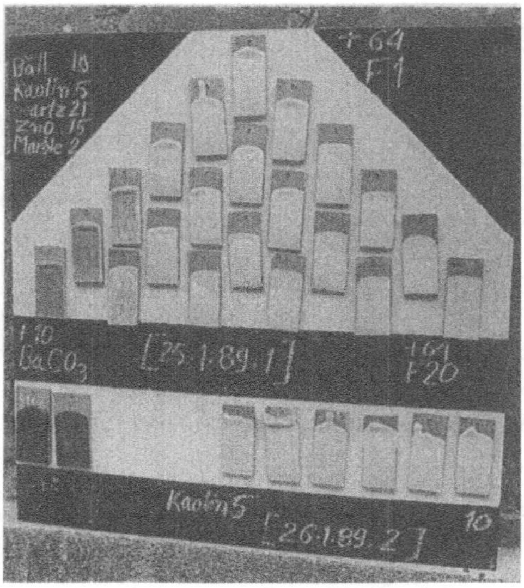

FIGURE 12.3.4.A The results of a 21-piece triaxial test are displayed on a board for easy comparison with other tests during ongoing research. The tiles hang on small nails and will later be stored on a string. Recipes are marked with chalk. A line blend is displayed at the bottom.

Glaze Test Record

name: . . . . test no: . . . .

prepared by : . . . . date : . . . .

Material	1	2	3	4	5	6	7	8	9
Maturity									
Surface									
Viscosity									
Crawling									
Crazing									
Colour									

Firing temperature : . . . . . °C Cone: . . . . .

Conclusion : . . . .

. . . .

FIGURE 12.3.4.B Example of a glaze test record form.

nally write down your conclusion like "make 1 kg test batch", "test again with 5% increase of clay". You could make a standard record form like the one in Fig. 12.3.4.B.

Testing is costly and the records help you to avoid unnecessary tests. When planning your next test, first take a look at your earlier results and compare them with your notes. When deciding which materials to add or which to decrease you may check with the oxide list below (page 122). As long as you work with one particular problem or line of research keep the test tiles close by for easy reference. Once you have finished the research you can store all the tests together by hanging them on a string in chronological order.

12.4. DEVELOPING A BASE GLAZE

The first step in developing a new glaze is to develop a base glaze, which is simply the combination of materials that melts at the desired temperature (without addition of colorants). Here, we describe an approach to making base glazes without knowing anything about the chemical composition of the materials.

Since all glazes require flux, stabilizer and glass former, these three materials are the starting point. There are a large number of fluxes available (divided into *primary* and *secondary* fluxes), but the stabilizer is usually china clay (kaolin) and the glass former is usually quartz (silica). The main differences are between low temperature and high temperature glazes. Below only the main materials (not chemicals) are mentioned. This list is a rough guide only.

Low temperature (900-1100°C)

Low temperature glazes require more flux and stronger flux than high temperature ones.

-- **primary flux**:	red lead, white lead, borax or boric acid, soda ash, gerstley borate (calcium borate), or most often frit (either lead, borax, or lead borosilicate).
-- **secondary flux**:	zinc oxide, barium carbonate, limestone, marble dust, talc.
-- **stabilizer**:	china clay or other clay.
-- **glass former**:	quartz, window glass powder.

High temperature (1100-1300°C)

-- **primary flux**:	feldspar, nepheline syenite, fusible clay, wood ash.
-- **secondary flux**:	zinc oxide, barium carbonate, limestone, marble dust, talc.
-- **stabilizer**:	china clay or other clay.
-- **glass former**:	quartz.

In the appendix there is a selection of glazes that can be used as a starting point for developing new glazes.

12.4.1. SELECTION OF MATERIALS

Materials necessarily have to be selected from what is available in your area, as most potters do not have access to suppliers with everything on hand.

When selecting materials to use in glazes, a general rule is to use materials that supply more than one oxide. For example, if magnesia (MgO) and silica (SiO_2) are both required, it is better to use talc ($3MgO \cdot 4SiO_2$) than magnesium carbonate and quartz. This is because the elements are already combined and contribute to a better glaze melt.

The biggest trouble with glazes is not to develop a nice new glaze but to keep it nice. Most materials vary from batch to batch and some materials may not be in regular supply. Therefore try to base your basic glaze on materials that you can rely on. Chemical stores often have ceramic oxides, but in a chemically pure form that is always very expensive. Instead look for the natural mineral containing the same oxide.

12.4.2. USING GENERAL RECIPES

There are hundreds of books on ceramics, most of which have recipes for glazes. These are limited in their usefulness, as often the raw materials are not available or are different from what you have in your country. Most of the time, these recipes do not work as expected and require modification. Without knowing the chemical analysis of materials, it is still possible to develop good glazes, using standard ones as a starting point and then modifying them systematically using the methods below.

LINE BLEND	PARTS BY VOLUME										
TEST NO.:	A	B	C	D	E	F	G	H	I	J	K
GLAZE	10	9	8	7	6	5	4	3	2	1	0
GLAZE + 30% ZnO	0	1	2	3	4	5	6	7	8	9	10
ZnO% IN GLAZE TEST	0	3%	6%	9%	12%	15%	18%	21%	24%	27%	30%

12.4.3. TESTING 2, 3 OR MORE MATERIALS USING LINE OR TRIAXIAL BLENDS

Line blends are the best place to start. A recipe for a glaze is made up, and then one material is selected to test in a line blend. It is added in steps, starting with a small amount and working up to perhaps 50% of the total, as described in section 12.3.2. This will give a range that may produce interesting results.

For example, use the following recipe from Ali Sheriff, Tanzania, for an unfritted borax glaze:

Boric acid	30%
Potash feldspar	25
Quartz	15
Dolomite	20
Ball clay	10

You might decide to see the effect of adding zinc oxide to the glaze. As a start a 10-step line blend is usefull (see the table above).

From this line blend you will get a good idea of how zinc oxide works in your basic glaze. Try the same with some more materials that are available like talc, limestone and zircon. From these line blends you will have a general idea of the amount of oxides which can be added.

The next step could be to try 2 or more materials in a triaxial blend. You might decide to try zinc oxide and talc. In this case, one point of the triangle would be 100% glaze, another point zinc oxide and the third point talc.

12.4.4. EVALUATING AND CARRYING OUT TESTS

After you finish a test, the next step is to evaluate it and decide how to proceed. Usually there will be at least one result that looks promising and, if you are really lucky, you might get a usable result the first time. Usually the best result from the first test will be the basis for further tests.

For example, if your zinc oxide line blend showed an almost-good glaze with 6% zinc oxide, you might want to try another line blend with smaller variations below and above 6% (see table below).

LINE BLEND	PARTS BY VOLUME										
TEST NO.:	A	B	C	D	E	F	G	H	I	J	K
A: GLAZE + 3% ZnO	10	9	8	7	6	5	4	3	2	1	0
B: GLAZE + 9% ZnO	0	1	2	3	4	5	6	7	8	9	10
Zn% IN GLAZE TEST	3%	3.6%	4.2%	4.8%	5.4%	6.0%	6.6%	7.2%	7.8%	8.4%	9%

If this is still not satisfactory, you might take the best result as the new base glaze and try to improve it in a new line blend, using another raw material. When deciding which materials to try, study the oxide list (page 122). Under each oxide you will find a list of its effects and you then choose accordingly. If your glaze is too stiff (high viscosity) you look for materials with low viscosity etc.

12.5. MODIFYING A BASE GLAZE

12.5.1. MATT GLAZE

Matt glazes have non-reflecting, dull surfaces, like eggshell, paper or river rocks (page 35). This kind of surface is called "matt". Matt glazes are especially popular for decorative ware, and for floor tiles because they are not slippery.

Matt glazes are developed in several different ways:

Underfired matt glaze

Most glazes that are fired below their maturing point become matt. In a similar way, overloading the glaze with a glaze material will produce a matt surface, because the material will act as a refractory that cannot be dissolved in the glaze melt.

– alumina matt: The addition of kaolin will produce a rather dull matt, but above 1200°C a smooth, pleasing matt is possible.

– silica matt: Excess amount of silica will cause small silica crystals to settle out of the melt during cooling. Alumina content should be low. If silica content is too high, the glaze will be matt from underfiring.

Crystalline matt glaze

During *slow* cooling, the glaze develops small crystals on the surface, which break up light and appear matt. These glazes are usually smoother than underfired matt glazes. If cooling is too rapid, crystals may not have time to develop, and the glaze will be glossy.

– barium matt: Barium carbonate is a common material to produce matt glazes, usually in amounts of 15-40 %. It is almost impossible to achieve a transparent matt glaze, but with luck it can be done with barium carbonate. Barium matt glazes are sensitive to firing conditions and it is better used together with other matting agents like zinc oxide and titanium dioxide.

– zinc matt: For low temperatures zinc oxide is a reliable agent for matt glaze. At temperatures above 1150°C it tends to build too large crystals, but a high alumina (Al_2O_3) content will reduce the size of the crystals. Pure zinc matt glazes are soft and not acid-proof, so for dinnerware it should be used in combination with other matting agents.

– titanium matt: Addition of 8-15% titanium dioxide will make a transparent glaze matt. The oxide easily combines with any iron in the body producing yellow to brown colors.

– calcium matt: The range of addition is 10-30% whiting ($CaCO_3$) or 20-40% wollastonite ($CaO \cdot SiO_2$). Bone ash ($Ca_3(PO_4)_2$) will produce smooth matt glazes for low temperatures when added to the frit.

– magnesium matt: Magnesium carbonate (magnesite $MgCO_3$), talc ($3MgO \cdot 4SiO_2 \cdot H_2O$) 10-18%, dolomite ($CaCO_3 \cdot MgCO_3$) often produce smooth, "buttery" matt glazes above 1100° C.

With a high amount of matting agent, the surface may turn too dull matt. This can be countered by either adding clay (alumina) that will reduce the crystal size or by reducing the matting agent.

Combining matting agents

A combination of matting agents will produce matt glazes less sensitive to firing conditions, harder and with better acid resistance. Below recipes of four different mixtures are suggested. The materials are premixed and added together to the glaze in amounts of 10-30% :

– Zinc oxide 50
  Kaolin 50
  Mixed and calcined above 800° C.

– Titanium dioxide 40
  Whiting 30
  Zinc oxide 30

– Titanium dioxide 30
  Tin oxide 30
  Zinc oxide 30

– Barium carbonate 40
  Whiting 20
  Zinc oxide 20
  Talc 20

The different mixtures are added to the glaze in line blends.

12.5.2. OPAQUE GLAZE

"Opaque" means you cannot see through the glaze (page 32). Opacity is developed by:

Opacifiers

These are finely ground materials that do not enter the glaze melt but remain as small white particles suspended throughout the glaze. They reflect light and make the glaze opaque. Standard opacifiers are:

- Tin oxide (SnO_2), addition 3-10%. Tin oxide is very expensive and is hardly used in the ceramics industry. It works well in combination with other opacifiers and produces a soft white color.

- Zircon (zirconium silicate, $ZrSiO_4$) is the main opacifier, addition 10-30%. It is used instead of the more expensive zirconium oxide (ZrO_2). Soda and potash content should be low. Very fine grinding promotes opacity. Commercial opacifiers are normally extremely finely ground zircon. It is better to add the zircon to the frit, but this may not be practical (see page 129).

- Titanium dioxide (TiO_2), addition 5-10%. Produces a creamish color and combines easily with iron in the body. Works well in combination with oxides of zinc, calcium and magnesium, especially in boron glazes. Opacifying effect depends on crystals forming during cooling.

- Bone ash (calcium phosphate, $Ca_3(PO_4)_2$), addition 5-15%. In amounts above 5% it may cause blistering and crawling in low temperature glazes and it is better added to the frit.

A variety of combinations of zinc, calcium, magnesium and titanium dioxide produces opacity in boron glazes. Zircon may be added (5-10%) to increase opacity further. By such combinations it is possible to produce a reliable zircon-based opaque glaze without the pinholing trouble otherwise seen with zircon glazes.

12.5.3. CRYSTAL, CRACKLE GLAZE

Crystal (or crystalline) and crackle glazes are used for special effects.

Crystalline glazes
Crystals develop in glazes that are low in alumina and that are cooled slowly. Usually these are small crystals that produce matt glazes.

FIGURE 12.5.3.A Glaze with decorative zinc crystals which grew during slow cooling.

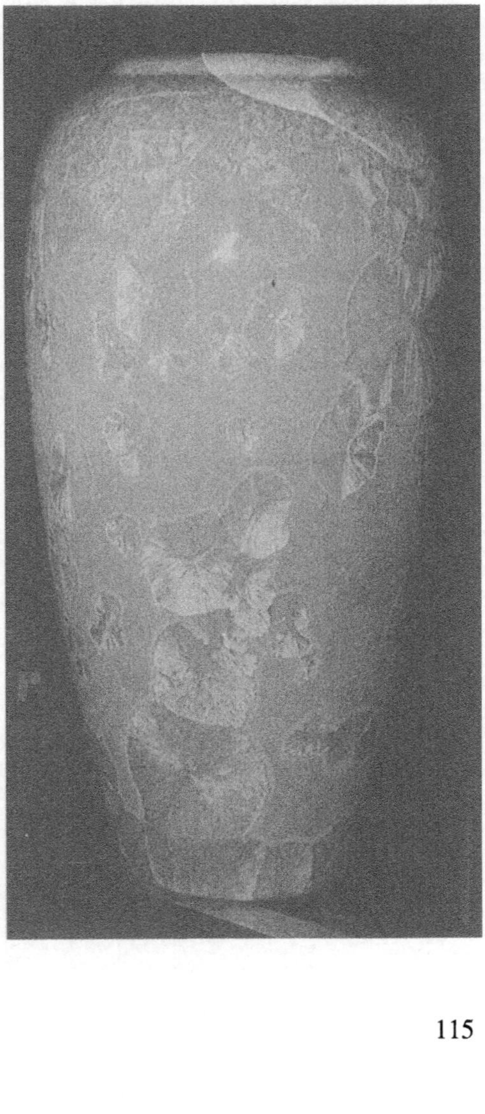

Very large crystals, from a few mm to several cm long, can be formed in special glazes. These glazes are fired to their maturing point, soaked for several hours and then cooled very slowly. That gives the crystals time to grow. To further increase the size of the crystals, the temperature can be kept slightly below the glaze's maturing point for several more hours. The outcome is very uncertain and many test firings are needed before the right firing and cooling method is developed.

Large crystals only grow in a very fluid glaze melt. So the glaze should contain little alumina and little silica but a large amount of flux. The best fluxes are lead, lithium, soda and potash.

The main agents for crystal formation are zinc oxide (20-30%) and titanium dioxide (5-15%). Lithium, calcium, magnesium and barium are supportive additions.

FIGURE 12.5.3.B Crackle transparent glaze on a white engobe with underglaze color. The crazing lines have been enhanced with ink.

Crackle glazes

These are glazes that craze, which are popular for decorative pottery. Crackle glaze should not be used on pots for food.

Most glazes can be made to craze by decreasing the quartz or increasing high-expansion oxides like soda and potash (see page 95 on crazing). Rapid cooling of the kiln helps to produce fine patterns of crazing.

To enhance the crackle, pots can be soaked in strong tea, or ink can be rubbed into the lines. Reglazing and refiring crackled pots with a contrasting glaze sometimes result in interesting patterns.

12.6. COLORED GLAZE

Colored glazes are developed by adding coloring oxides. These are added to the base glaze as a percentage, based on the range for each oxide as listed below. Different oxides have different strengths, so some of them are used in much larger amounts than others.

For example, you might want a brown glaze. Looking at the list of oxides, you find that brown can be developed with iron oxide from 5-10%. This can be done as a line blend, adding 5,6,7,8,9 and 10% to the base glaze. The percentage is in addition to the total base glaze weight:

Glaze	100 g
Iron oxide	6%

$100 \text{ g} \times 0.06 = 6 \text{ g}$

Glaze + oxide = 106 g

Ready-made glaze pigments, called glaze stains, are also used to develop colors that cannot be made easily with oxides alone.

12.6.1. LIST OF OXIDE ADDITIONS

It is more or less impossible to give an accurate guide to colors in glaze, because there are so many variables of chemical reaction in different base glazes.

The firing conditions, temperature and oxidation/reduction also greatly influence the color of the glaze.

The table below should be considered a rough guide. See also chapter 13 (page 122) for color reactions in different types of base glazes.

12.6.2. LINE, TRIAXIAL BLEND PLANNING

The most interesting colors often come from combining 2 or more oxides in the same base glaze. Usually it is best to test the base glaze first with various oxides alone and to use the best results in combination with each other. Line blends are useful for this kind of test, and triaxial blends can also be used for 3 oxides in combination (see page 105, 107 for details on line and triaxial blending).

One-color line blend

For testing a color oxide, you prepare two mixtures for a line blend.
Example:
Mixture A: 100 parts your basic glaze
Mixture B: 100 parts basic glaze
10 parts titanium dioxide

Make line blends with all the coloring oxides you have. After firing, you will

Single oxides	Percent	Effects
Iron oxide, Fe_2O_3	1 – 5%	Green, cream, light brown
	5 – 10%	Brown, red-brown
	10 – 15%	Dark brown, black
Cobalt oxide, CoO Cobalt carbonate, $CoCO_3$	0.2 – 3%	Blue
Manganese dioxide, MnO_2 Manganese carbonate, $MnCO_3$	2 – 10%	Brown, purple-brown
Rutile, TiO_2	1 – 10%	Yellow, tan, mottled colors
Chrome oxide, Cr_2O_3	1 – 5%	Green
Copper oxide, CuO Copper carbonate, $CuCO_3$	0.5 – 5%	Green, blue, red in reduction
Nickel oxide, NiO	0.5 – 3%	Gray, green-brown
Ilmenite, magnetite (contains iron)	1 – 10%	In granular form produces specks and spots in the glaze.
Antimony oxide, Sb_2O_3	1 – 5%	Cream to yellow in lead glazes

have a good idea of the color range you can get with your basic glazes. Maybe you will already now have all the colors you need. If you want to try a combination of several oxides you can do this by line blends or triaxial blends.

Two-color line blend

Choose one of the colors you got from your first set of line blend testing. Make this your basic glaze and then try another coloring oxide in addition to this.

Example:

Mixture A: 100 glaze

Mixture B: 100 glaze
4 copper oxide
5 iron oxide

Note that when mixing several coloring oxides their total amount should normally not exceed 10% of the glaze.

This type of line blending can be continued with any combination of oxides. Do it one step at a time with only one or two line blends at a time in your regular glaze firing. After firing you can choose the best results and do more tests along those lines.

Triaxial blend

From your first set of line blends choose three coloring oxides and test their combinations in a triaxial blend. When setting up the triaxial blend, make the points A, B and C with oxide additions about 30% higher than what you expect to use in the final glaze.

You can even try four color oxides in one triaxial blend.

Example:

Your line blend showed that 1.5% addition of cobalt oxide produced a nice blue, but you want to modify it with other color oxides:

Base glaze: glaze + 1.5% cobalt oxide
A: base glaze + 6% iron oxide
B: base glaze + 5% copper oxide
C: base glaze + 8% titanium dioxide

After doing the tests you have to calculate the final recipe. This is done by setting up a calculation table as shown on page 109.

12.6.3. COLOR PIGMENTS

Glazes can be colored by adding metallic oxides directly to them. Some oxides can be used as on-glaze colorants by painting them directly on the unfired glazed object.

Ceramic pigments are produced from the same coloring oxides, but other materials are added in order to change the colors and make them more stable or cheaper.

The materials used for pigments can be divided into four groups:

Color agent

- metallic oxides; for example, iron oxide, copper oxide.

Modifier

- influences coloring effect of oxides. Examples of modifiers: titanium dioxide, zinc oxide, zirconium oxide, antimony oxide.

Filler

- raises melting point of the pigment and stabilizes the coloring oxides. Examples of fillers: alumina, quartz, feldspar, clay body.

Flux

- lowers melting point of the pigment. Examples of fluxes: borax, lead, frit or glaze.

Fluxes are added according to the use of the color pigment. The pigments can be adjusted for use as:

Under-glaze colorant:

- The pigment is painted directly on the raw or biscuit-fired body and a glaze is applied on top.

Maiolica or on-glaze:

- Decoration on the unfired glaze layer.

Overglaze enamel:

- Applied to the already fired glaze.

In-glaze colorant:

- Added to a basic glaze as a coloring agent.

Production of Color pigments

Close production control, accurate weighing and the use of the right materials are especially important when producing color pigments. Even slight deviations may result in the change of a fired colour.

Four main processes are used in the production:

1) Mixing of raw materials
2) Calcination
3) Washing
4) Grinding

Mixing

If all raw materials of the recipe are already finely ground mixing can be done manually ensuring good mixing by screening the batch twice through 60 mesh. Normally materials will be coarse, so after weighing out the pigment recipe the batch is ball-milled.

After milling drying and calcination follow.

Calcination

The calcination will burn away carbonates, water, sulfates and the coloring oxides will form new crystalline combinations with the other materials in the batch. This will stabilize the colors so that they will not be easily dissolved in the glaze.

The temperature of calcination is in the range of 700°C to 1400°C. In general, the color pigment should be calcined at least to the temperature at which it is going to be used and preferably higher. Some colors will disappear if fired high whereas other colors will only develop correctly at 1300°-1400°C.

Calcination is done in small saggars or clay pots with a lid. The pigments are fired in a small kiln (e.g. test kiln) to the desired temperature or in the hot spots of the normal production kiln.

Washing

After calcination the sintered pigments are crushed to sand size and then washed with water in order to remove any soluble materials that may remain. The washing is normally not important except for pigments to be used in delicate decorations where possible soluble materials may cause a blurred final image.

Grinding

The pigment is ground in a small ball mill. For enamel overglaze decorations it should be ground very fine. In normal practice it should pass 250 mesh. When used as a glaze colorant, 150 mesh is fine enuough, but in general the coloring quality is better with fineness.

For special decorative speckled effects the pigment can be made coarse grained.

After grinding the pigment is dried, packed and labeled and a color test made before releasing for sale or production.

The basic pigment can now be used for mixing of underglaze, on-glaze or enamel colorants with additions of fluxes, clay, silica etc. as described below

Underglaze

These colorants are applied to raw body, body covered with engobe or to biscuit-fired body. Colored engobes can also be termed underglaze colors.

The colorants should not react with or be dissolved by the overlying glaze. A high content of clay, feldspar or whiting prevents this.

If applying to raw clay, shrinkage should be adjusted to fit with that of the body. For biscuit body some 5-10% raw clay will give better adhesion and strength to the dried surface. 3-5% raw borax reduces tendency of glaze crawling over the decoration and adds strength to the decoration before glazing.

FIGURE 12.6.3.A The transparent glaze crawled and lifted off the underlying decoration which did not stick to the body.

10-20% addition of the glaze used for final glazing is normally also added.

Addition of glue like sugar, dextrin, CMC helps adhesion.

Maiolica or on-glaze

These colorants are applied onto the already glazed but unfired pot. The colorants sink into the glaze during firing and melt together with the main glaze. More fluxes are added to maiolica colorants than to underglaze colorants and the lower the viscosity of the colors and the glaze is, the more the decoration will run and the contours of the decoration will be blurred.

About one part frit is added to one part pigment. With a low melting frit or a pigment containing a high amount of copper oxide the frit content is lowered.

Maiolica colorants can be made by adding a little glaze or frit to the raw color oxide. The maiolica technique can also be used by decorating with coloured glazes on top of the basic glaze. To prevent running, the melting point of the colored glaze can be raised by adding silica and clay. Color oxide mixed with water can also be used when thinly applied. The oxide will then melt together with the glaze. If the oxide layer is too thick the glaze cannot "wet" the oxide and the decoration will be dry and dark in color after firing.

Overglaze enamel

Overglazes (or "China paints") consist of frit and pigment and they are fired at low temperatures of 700°-850°C. The flux content is 70-90% of the enamel color.

Examples of lead-free fluxes for 400-600°C

1)	Borax	380	2)	ZnO	37
	Quartz	100		Borax	60
				Whiting	7

The flux and the color pigment are melted together and ground.

The ground colorant is mixed with about 50% organic oil (linseed oil, olive oil) as a medium for painting on the fired glaze surface. Turpentine is used for thinning. If no proper oil is available turpentine which has had some of its volatile parts removed by boiling can be used as a medium. Another medium for suspending the colorant is water with the addition of white carpenter's glue.

Glaze colorant

The color pigments can also be used for coloring basic opaque or transparent glazes. Coloring can be done by directly adding color oxides to the glaze. However, there are some benefits from doing the coloring with prepared pigments:

- The color effect of oxides is increased and thus cost of expensive oxides like cobalt can be reduced.
- Colors can be made more stable so they will be less influenced by kiln atmosphere and glaze materials.
- More colors can be produced.
- Blisters and pinholes produced by coloring oxides (MnO) can be avoided.

13. GLAZE OXIDES

This list of glaze materials only includes materials that we can expect to obtain easily. It also lists mineral sources for the oxides, which may be helpful when consulting with geologists. The formula suggestions are only meant as a rough guide for those who work with formulas.

The following list can be used as a reference when developing or modifying glazes. If you have a problem with pinholes, then go through the list noting down all materials that are mentioned as having high viscosity or high surface tension or the opposite. Comparing your glaze recipe with your notes will give ideas of what materials to increase or decrease.

"MP" means "melting point".

ALUMINUM OXIDE, ALUMINA,
Al_2O_3 (Stabilizer), MP 2050°C

Source:
- Aluminum oxide, alumina, (Al_2O_3)
- Clay ($Al_2O_3 \cdot 2SiO_2 \cdot 2H_2O$)
- Feldspar ($K_2O/N_2O/CaO \cdot Al_2O_3 \cdot 6SiO_2$)

Mineral sources:
Kaolin, ball clay, bentonite, corundum, bauxite, silimanite, kyanite, gibbsite (hydrargillite), websterite (aluminite), alunogen.

Effect:
- Increases melting point, hardness, viscosity, surface tension.
- Reduces tendency of crystal formation.
- Reduces thermal expansion.
- Small additions help other opacifiers.
- Large amounts produce matt glazes.

Clay addition normally 5-15%. Clay helps to keep glaze materials suspended in the bucket. Large additions cause problems of cracking of raw glaze layer and crawling, pinholing.

Formula:
- Ratio in shiny glazes,
  Alumina : Silica = 1:6-1:10
- Ratio in matt glazes,
  Alumina : Silica = 1:4-1:2

Color effect:
- Increase of alumina makes red and yellow-brown iron oxide color more brownish.
- MnO colors turn brown and less violet.
- CoO colors turn darker.

BARIUM OXIDE, BARIA,
BaO (flux), MP 1923°C

Source:
- Barium carbonate ($BaCO_3$). Poisonous if it enters the blood.
- Barium sulfate ($BaSO_4$)
- Selenite ($BaO \cdot SeO_2$)

Mineral sources:

Witherite, barytes, celsian, bromlite, barytocalcite.

Effect:

- BaO reduces boron's tendency to form opaque "clouds" and therefore helps to make boron glaze transparent.
- Reduces chemical resistance.
- High amounts (above 25%) produce matt glaze due to formation of crystals. BaO matt glazes are not stable.
- Lowers melting point.
- Slow in giving off CO_2. Sometimes sulfate problems in coal- or oil-fired kilns.
- Helps formation of crystalline glazes.
- Improves hardness.
- Small amounts improve gloss.

Formula:

- Generally, below 1100°C BaO should be less than 0.10 mole.
- Above 0.3 mole BaO raises melting point of glaze.

Color effect:

- CoO colors turn more violet, Cr_2O_3 below 1% turns more yellow.
- CuO colors turn from green to blue-green.
- Iron colors are subdued.
- NiO colors turn more brownish.

BORIC OXIDE,

B_2O_3 (stabilizer or glass former), MP 741°C

Source:

- Borax ($Na_2B_4O_7 \cdot 10H_2O$)
- Boric acid ($B_2O_3 \cdot 3H_2O$)
- Both materials are soluble in water and they are normally introduced in a frit.
- Colemanite, gerstley borate ($2CaO \cdot 3B_2O_3 \cdot 5H_2O$). The only insoluble mineral form of borax, only mined in the USA.
- Calcium borate ($CaO \cdot B_2O_3 \cdot 6H_2O_2$), the chemical form of colemanite.

Boric oxide is sometimes classified as a stabilizer (USA) and sometimes as a glass former (UK).

Mineral sources:

Borax (tincal), kernite, ulexite, colemanite, boracite, sassolin.

Effect:

- Strongly lowers melting point. Mainly used below 1100°C.
- Improves formation of an intermediate layer between glaze and body.
- Boric oxide below 15% reduces tendency to craze, higher amounts increase crazing.
- Lowers viscosity and surface tension.
- Low thermal expansion rate.
- B_2O_3 less than 10% lowers surface tension.
- High content of boric oxide forms opaque clouds especially in combinations with CaO and SnO_2. This is reduced by addition of BaO or $SrCO_3$.
- Extends the firing range.
- Reduces tendency to crystallize.

Formula:

- Boric oxide ratio to silica is normally 1:10 and should not be less than 1:2. In frits a ratio below 1:2 will leave the frit water-soluble.

Color effect:

- MnO colors turn a violet hue.
- Iron colors become yellowish-reddish.

- CoO colors become brighter.
- CuO colors change from green to bluish green.

Sometimes a small percentage of raw borax is added to glaze or to engobe. When the glaze layer dries, the borax recrystallizes and this gives strength to the raw glaze layer which means it will not be damaged during handling.

CALCIUM OXIDE, CALCIA,

CaO (flux), MP 2570°C

Source:

- Calcium carbonate ($CaCO_3$), limestone, whiting, marble.
- Wollastonite ($CaO \cdot SiO_2$).
- Dolomite ($CaCO_3 \cdot MgCO_3$).
- Anorthite, lime feldspar ($CaO \cdot Al_2O_3 \cdot 2SiO_2$).
- Calcium sulfate ($CaSO_4$), plaster of paris.
- Calcium borate ($2CaO \cdot 3B_2O_3 \cdot 5H_2O$).
- Calcium fluoride (CaF_2).
- Calcium phosphate ($Ca_3(PO_4)_2$) (bone ash).

Mineral sources:

Glauberite, fluorspar, apatite, lime, calcite, chalk, limestone, marble, gypsum, alabaster, seashells, coral, portland cement.

Effect:

- Combines readily with silica in glaze and, if CaO is present in body, it reacts with SiO_2 in glaze to form a strong interface, reducing crazing.
- Increases hardness, especially with boron glazes.
- Reduces tendency to craze.
- Primary flux for temperatures above 1100°C.
- Below 1100°C small additions act as secondary flux.
- High CaO produces opacity in boron glazes, and white matt wax-like glazes can be produced.
- Too high CaO gives dull, matt finish.
- $CaCO_3$ gives off CO_2 at 825° C.
- In zircon white glaze CaO increases pinholes and a dull surface.
- Decreases lead solubility.

Formula:

- At cone 03 CaO not above 0.25-0.28 mole
- At cone 01 CaO not above 0.30-0.35 mole

Color effect:

- CaO turns Cr_2O_3 colors yellow.
- MnO browns and violets are improved with CaO.
- CaO is important for production of iron-red, chrome-green and blue color pigments.

LEAD OXIDE,

PbO (flux), MP 888°C

Source:

- Litharge (PbO)
- Red lead (Pb_3O_4)
- White lead, lead carbonate ($2PbCO_3Pb(OH)_2$)

Lead is a very good flux but it is very poisonous and expensive. It should never be used in ware that will contain food, but still is used frequently for decorative ware. If you use lead, it should always be in frit form.

Mineral sources:

Cerussite (white lead), galena (blue lead), minium (red lead), anglesite, pyromorphite.

Effect:
- Smooth, shiny low-temperature glazes.
- Strong flux.
- Good for transparent glazes.
- Reduces viscosity and surface tension.
- Reduces hardness and chemical resistance.
- Evaporates easily during firing.
- Combined with boric oxide, it is a common flux for earthenware glazes.
- It is more dangerous with copper oxide, which increases lead release 10 times.
- Small amounts in high temperature increase smoothness.

Formula:
Simple lead-alumina-silicate combinations make glazes in the following ratios[3]:

$PbO \cdot 0.10\ Al_2O_3 \cdot 1.0\ SiO_2$	900°C
$PbO \cdot 0.11\ Al_2O_3 \cdot 1.1\ SiO_2$	920°C
$PbO \cdot 0.12\ Al_2O_3 \cdot 1.2\ SiO_2$	940°C
$PbO \cdot 0.13\ Al_2O_3 \cdot 1.3\ SiO_2$	960°C
$PbO \cdot 0.14\ Al_2O_3 \cdot 1.4\ SiO_2$	980°C
$PbO \cdot 0.15\ Al_2O_3 \cdot 1.5\ SiO_2$	1000°C
.....	
$PbO \cdot 0.25\ Al_2O_3 \cdot 2.5\ SiO_2$	1200°C

Color effect:
- Good with almost all colorants.
- Lead transparent glazes produce pleasant colors for engobe decorations.
- With iron, rich tans, browns, reds.
- With copper, rich greens (caution: lead release is increased 10 times).
- With antimony oxide, yellow.

[3] W. Lehnhauser, *Glasuren und ihre Farben*, p. 16.

LITHIUM OXIDE,
Li_2O (flux), MP above 618°C

Source:
- Lepidolite (lithium mica) 1.5%-6% lithium oxide.
- Petalite ($Li_2O \cdot Al_2O_3 \cdot SiO_2$), 2%-4% lithium oxide.
- Spodumene ($Li_2O \cdot Al_2O_3 \cdot 4SiO_2$), about 8% lithium oxide.
- Lithium carbonate (Li_2CO_3).
- A number of artificial lithium chemicals exist.
- High price.

Effect:
- A strong flux
- Lowers viscosity.
- Improves hardness.
- Improves gloss.
- High Li_2O content furthers formation of crystals in the melted glaze.
- Already 1% additions of Li_2CO_3 improve gloss and smoothness of glaze.

Color effect:
- CuO turns to blue colors.
- In lithium glaze 1% SnO_2 + 0.5% CuO produces Chinese reds in reduction firings.

MAGNESIUM OXIDE, MAGNESIA,
MgO (flux), MP 2800°C

Source:
- Talc ($3MgO \cdot 4SiO_2 \cdot H_2O$).
- Magnesite (magnesium carbonate) ($MgCO_3$)
- Dolomite ($CaCO_3 \cdot MgCO_3$)

Mineral sources:

Soapstone or steatite, serpentine, meerschaum, vermiculite, periclase magnesia, magnesite, brucite.

Effect:

- Raises melting point.
- High surface tension.
- Reduces crazing due to its low thermal expansion.
- Small amounts increase gloss.
- Larger amounts make matt glaze (best above 1100°C).
- With double glazing, good for special-effect crawling glaze.

Formula:

- Below 1100°C, less than 0.1 mole MgO increases gloss and 0.2 - 0.4 mole MgO produces matt glazes.

Color effect:

- CoO blue turns violet with MgO.
- MgO glaze on red iron rich body turns the red color to a dirty yellow-brown color. Therefore transparent glaze should contain no MgO.
- Cr_2O_3 green only accepts small amounts of MgO. Large amounts bleach the green color.

PHOSPHORUS OXIDE,

P_2O_5 (glass former), MP 569°C

Source:

- Bone ash, calcium phosphate ($Ca_3(PO_4)_2$)
- Apatite, $3Ca_3(PO_4)_2Ca(Cl\ F)_2$.

Mineral sources:

Bone ash (made from calcining animal bones), apatite, wavellite, vivianite.

Effect:

- P_2O_5 can replace some of the SiO_2 in the glaze.
- Strong flux, especially with MgO, BaO and alkalis.
- Additions above 5% form opaque glaze, especially in combination with ZnO and in lead-free glazes.
- Additions of up to 4% may increase melting and reduce pinholes. However, bone ash often increases pinholes due to high release of gas (instead add the bone ash to the frit).
- High additions (above 10%) produce matt glaze.
- Additions above 25%-30% make the glaze too soluble (less acid- or weather-resistant).

Color effect:

- CoO blue turns more violet.
- In B_2O_3 glazes iron colors turn yellowish.
- In alkaline glazes iron colors turn white with high amount of P_2O_5.
- CuO greens turn bluish and with high P_2O_5 spotted.
- MnO colors turn more violet.
- Cr_2O_3 colors are improved to lighter shades.
- Interesting special surface effects with high P_2O_5.

POTASSIUM OXIDE, POTASH,

K_2O (flux), MP 896°C

Source:

- Potassium carbonate, potash (pearl ash) (K_2CO_3), water-soluble.
- Potassium nitrate, saltpeter (KNO_3), water-soluble – also used as fertilizer.

- Potash feldspar ($K_2O \cdot Al_2O_3 \cdot 6SiO_2$), exists as minerals named orthoclase and microcline, melting at 1200°C.
- Nepheline syenite ($3Na_2O \cdot K_2O \cdot 4Al_2O_3 \cdot 8SiO_2$).

Mineral sources:

Saltpeter, potassium bichromate, leucite

Effect:

- Potash's effect is very similar to soda, but it is a slightly less powerful flux.
- Potash increases crazing, but a little less than soda does.

SILICON OXIDE, SILICA,
SiO_2 (glass former), MP 1710°C

Source:

- Quartz, SiO_2
- Clay, $Al_2O_3 \cdot 2SiO_2 \cdot 2H_2O$
- Feldspar, $Na_2O/K_2O/CaO \cdot Al_2O_3 \cdot 6SiO_2$
- Talc, $3MgO \cdot 4SiO_2 \cdot H_2O$
- Zirconium silicate, $ZrSiO_4$
- Wollastonite, $CaO \cdot SiO_2$

Mineral sources:

Flint, chalcedony, chert, sand, quartzite, diatomite, granite, part of all rocks.

Effect:

- A glass former, a part of all glazes.
- Generally raises melting temperature.
- Low thermal expansion, addition reduces crazing.
- Addition to body also reduces crazing (see glaze faults).
- Increases viscosity of glaze melt.
- Increases acid and weather resistance.
- Increases hardness of glaze.
- High amounts make the glaze shiver.

Formula:

- Addition of 0.1 mole SiO_2 increases melting point by 20°C.
- Amount of SiO_2 depends on other glass-forming oxides. In general, earthenware 1-2.5 mole SiO_2 and stoneware 1-4 mole SiO_2.

Color effects:

- SiO_2 has little influence on effect of coloring oxides.

SODIUM OXIDE, SODA,
Na_2O (flux), MP about 800°C

Source:

- Sodium carbonate (Na_2CO_3) as crystal soda or calcined soda – also named soda ash, soluble in water, absorbs moisture from the air.
- Sodium nitrate ($NaNO_3$). Sodium saltpeter (Chile saltpeter), soluble in water.
- Sodium chloride (NaCl). Table salt, water-soluble, used in salt glazing, used in frit for reducing discoloration of frit by iron compounds.
- Soda feldspar or albite ($Na_2O \cdot Al_2O \cdot 6SiO_2$), a white mineral melting at 1170°C.
- Nepheline syenite, ($K_2O \cdot 3Na_2O \cdot 4Al_2O_3 \cdot 8SiO_2$), mineral melting at 1100°- 1200°C.

Mineral sources:

Natron, halite, hauynite, plagioclase,oligoclase, sodalite, glauberite, cryolite, glauber salt.

Effect:

- Strong fluxing agent.
- Improves gloss.

- Very high thermal expansion induces crazing.
- Lowers elasticity of glaze, which becomes brittle with high amount of Na_2O.
- Low viscosity, causes glaze to run.
- Short melting range.
- Evaporates easily above 1100°C (salt glazing).

Color effect:
- High amount of Na_2O or K_2O produces "alkaline colors", noted for their brightness and interesting shades.
- Copper oxide turns blue instead of green.
- Manganese oxide turns violet.
- Cobalt gives a light blue.
- Iron oxide produces red in connection with boron.

Formula:
- In alkaline frits 1 mole alkali with at least 2.5 mole SiO_2, otherwise the alkalis Na_2O and K_2O will remain water-soluble.

TIN OXIDE,
SnO_2 (glass former group), MP 1930°C

Source:
- Tin oxide, SnO_2 (artificial)

Mineral sources:
Cassiterite (tinstone), stannite, tin pyrites.

Effect:
- Opacifier with 5-10% addition, less efficient in alkali-rich glazes.
- Opacifying effect increases with CaO, TiO_2 and ZrO_2. Fine grinding improves opacifying effect.
- Increases viscosity and melting point.
- Increases hardness and acid resistance.
- Increases elasticity of glaze (reduces crazing)

Color effect:
- In leadless glaze turns CuO bluish.
- Produces pink in combination with Cr_2O_3 and CaO.
- Iron brown colors turn redder.
- Manganese brown turns more violet.
- Used for stabilizing colors in pigment production.

TITANIUM DIOXIDE
TiO_2 (glass former group), MP 1855°C

Source:
- Titanium dioxide, titania, TiO_2 (artificial)
- Rutile, TiO_2 (85-98% TiO_2)
- Perovskite, $CaO \cdot TiO_2$
- Titanite, $CaO \cdot TiO_2 \cdot SiO_2$
- Ilmenite, $FeO \cdot TiO_2$

Mineral sources:
Rutile, anatase, brookite, titanite (sphene), ilmenite.

Effect:
- Opacifier but not so reliable. Opacity improves with addition of ZnO and CaO.
- Above 10% TiO_2, glaze turns matt due to forming of small crystals if cooling is slow. Mattness depends very much on firing conditions.
- Reduces crazing.
- Increases acid resistance.
- Reduces lead solubility when introduced in small amounts.
- Used for crystal glazes in combination with ZnO.

Color effect:

- Pure TiO_2 produces white colors in alkali-rich, lead-free glazes.
- In lead glazes and high boron glazes with small amounts of iron oxide a slight yellow color is obtained.
- Rutile contains some iron. The pure TiO_2 will work as rutile with an addition of about 5% iron oxide.
- On iron-rich bodies (red firing) TiO_2 combines with the iron of the body to form yellow-brown colors.
- TiO_2 addition turns CoO blue to gray-blue and with high CoO to green.
- Low CuO turns yellowish, high CuO bluish.
- Cr_2O_3 becomes dirty grayish.
- MnO_2 turns grayish
- NiO red and blue colors changed to green.

ZINC OXIDE,
ZnO (flux), MP 1975°C

Source:

- Zinc oxide, zinc white, ZnO
- Zinc borate, $ZnO \cdot B_2O_3$
- Zinc chloride, $ZnCl_2$
- Zinc phosphate, $3ZnO \cdot P_2O_5$

Mineral sources:
Sphalerite or blende (zinc sulfide), the original zinc ore, smithsonite, hydrozincite, willemite.

Effect:

- Above 1100°C a strong flux.
- In small amounts increases brilliance.
- High amounts produce matt glazes.
- Reduces viscosity, increases surface tension.
- Increases boron clouds and helps opacity in combination with other opacifiers.
- Reduces crazing due to its low thermal expansion and high elasticity.
- Its high drying shrinkage may cause crawling if added without prior calcination.
- In high amounts best agent for forming crystals.
- Produces special surface and color effect in high boron glazes.

Color effect:

- Generally increases brightness of colors.
- Chrome-green turns gray.
- Cobalt blue becomes lighter with less of a violet hue.
- Manganese violet turns brown.

ZIRCONIUM OXIDE
ZrO2 (glass former group), MP 2700°C

Source:

- Zircon, zirconium silicate, $ZrSiO_4$
- Zirconium oxide, zirconia, ZrO_2
- Commercial opacifiers
- Commercial zircon frits.

Mineral sources:
Zircon is found in beach sands, baddeleyite (ZrO_2).

Effect:

- Zircon additions of 10-20% produce opaque white glaze (due to its high price zirconium oxide is seldom used).
- Used in combination with ZnO, MgO, BaO, SnO_2 opacity is increased.
- Opacity is furthered by fine grinding and by adding zircon to the frit instead of the batch.
- Increases melting point.
- Increases hardness, viscosity and surface tension.
- Increases tendency to form pinholes.
- Reduces crazing.

14. QUALITY CONTROL

A successful business depends on consistent results. This can only be done if quality control is made a habit. This means having regular procedures for storing glaze materials, checking new shipments, weighing, grinding, mixing, and checking each new batch of glaze before using it in production.

14.1. RAW MATERIALS CONTROL

Raw materials suppliers have their own problems with getting consistent materials. Sometimes they may send you a different material without any notification, or the quality of material from the mine may change. If you have enough working capital and storage area, it is best to get raw materials in large quantities, up to one year's need.

14.1.1. RAW MATERIAL TESTING

When you get a new shipment of raw materials, each one should be tested. For the individual potter this is simply done by testing each one in the standard glaze recipe to see if there is any change. About 200 g glaze is mixed using the old stock materials and replacing only one of the new materials at a time. If the test glaze is different from your standard glaze, it will be necessary to alter your glaze recipe.

14.1.2. STORING OF GLAZE MATERIALS

All materials should be kept in bags or buckets so there is no chance of mixing up different materials. Mark the contents of all bags and buckets and the material's delivery date on labels that cannot easily be removed. Keep the glaze material store separated from working areas and make sure that only responsible persons have access to it.

14.2. GLAZE PREPARATION CONTROL

Many glaze problems are caused by carelessness during mixing of the glaze. When preparing glazes and frit, make sure that the right recipe is used and that the weighing is done correctly.

Batch cards

If you are running a small pottery and you are doing all glaze work yourself, you can rely on a very simple system. Still, write down your recipe, keep it next to the balance and after weighing each material tick it off on the recipe.

For larger productions use a batch card system. A batch card form is shown in Fig. 8.3.0.B. The card follows the glaze batch during its preparation and later when the glaze is used in production. It has three purposes:

- It shows the glaze mixer the recipe, ball milling time, density of the glaze slip.
- The supervisor can easily check if all instructions are followed.
- If something goes wrong, the batch card helps to trace the cause of the problem.

The batch card number should be marked on the glaze bucket. To avoid mistakes tie a tile glazed with the same glaze to the bucket.

Balance
The balance and the weights need to be checked now and then. The weights should be clean. The balance may become inaccurate because the scales get dirty or the pivots or beams get out of alignment. After cleaning the weights they and the balance are checked by weighing something with a known weight (1 liter of water weighs 1 kg).

Graduated cylinder
Cylinders or flasks used for measuring volume are used for adjusting density of glaze slips. Unfortunately, measuring cylinders are often not graduated correctly by the manufacturer. The cylinder can be checked by filling it with water to its mark, say 250 ml and then checking if the water weighs 250 g. In some cases they have been out by more than 10%.

Ball milling

The fineness of the glaze particles influences the glaze very much. To keep this constant, make sure that the ball milling time is the same. The time should be noted on the batch card. If different glazes are milled in the same ball mill, the worker must enter on the card that he has cleaned the ball mill before loading it. The supervisor should check that the ball mill lining and pebbles are correct.

Sieving
The glaze should be screened before use. On the batch card screen mesh size is mentioned. Check the residue on the screen. If you get more residue than usual, there may be something wrong with the ball milling.

14.2.1. METHODS OF TESTING BATCHES OF GLAZE AND FRITS

Testing frit
Molten frit can be drawn from the frit kiln to see whether all ingredients are well melted and whether air bubbles are released. Air bubbles may not be a problem, since many of them will be released during grinding and the second glaze firing. But if air bubbles (pinholes) give trouble during glaze firing, it may be a good idea to extend fritting time, so that the air has time to escape. In continuous frit kilns, bars of refractory brick can be placed on the sloping floor to slow down the flow of frit.

After fritting is over, the melting temperature and the viscosity of the frit can be compared with previous batches of frit by melting a fixed amount of frit on a sloped tile.

Testing glaze
Each new batch of glaze should be made at least one firing before using it. This will give enough time to apply the glaze to a few

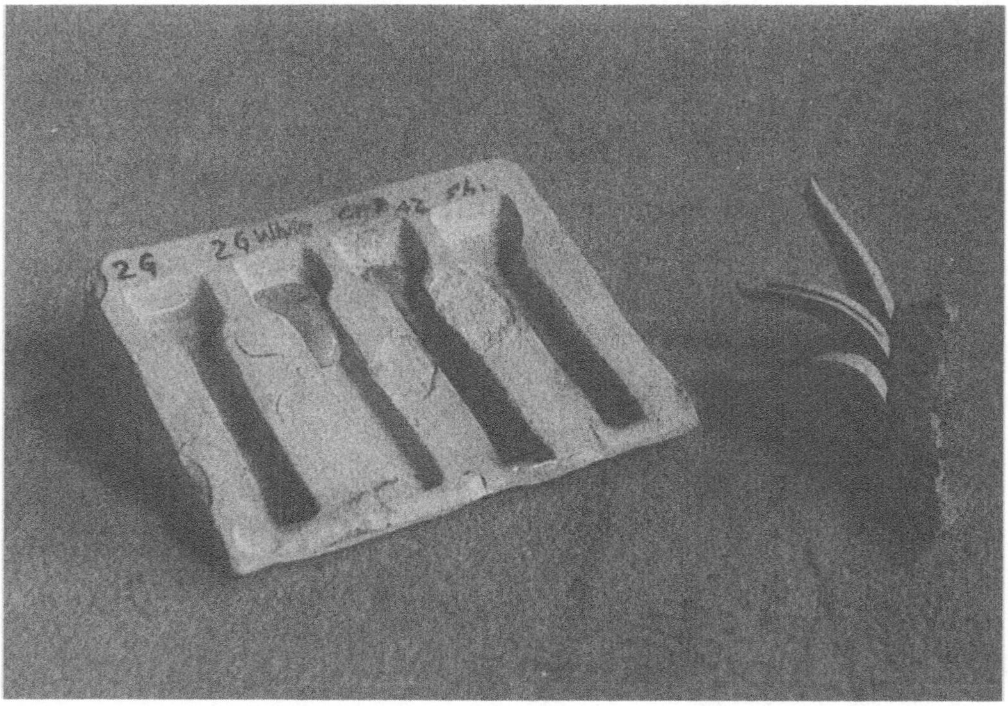

FIGURE 14.2.1.A Flow testing device for frit or glaze. Equal amount of test material is placed at top. For comparison, a new frit should always be tested along with a known frit.

test pieces and fire them in the regular glaze firing. Glaze at least three pieces and place one in a cold spot, one in a normal and one in a hot spot. If something is wrong with the glaze, this will prevent a whole kilnload from being ruined.

15. HEALTH AND SAFETY

As in all other types of industries, precautions are needed to avoid health hazards to the ceramics workers.

15.1. MACHINERY

Moving parts of machinery used in the workshop should be enclosed to prevent hands, clothing or hair being caught in them. The belts and gears of ball mills, hammer mills etc. are especially dangerous.

Place the electrical switch next to machines, where the operator can reach it.

FIGURE 15.2.0.A Grinding of frit in a plate mill. The worker must wear a dust mask.

15.2. DUST

Workers in the ceramics industry are constantly exposed to dust. Inhalation of dust from clay materials and quartz will cause silicosis. This is an incurable lung disease. The dangerous dust is so fine it cannot be seen.

The workshop floor should be cleaned regularly by scrubbing it with water. Dry sweeping should never take place. If it is not possible to wash the floors they can be swept after spreading wet or better still oiled sawdust. Tables, shelves and other surfaces collecting dust should be cleaned with a wet sponge at least once a week.

Dry blending of glaze and clay materials should be avoided. If it is done, the worker must wear a dust mask.

If the climate allows it, keep doors and windows open. Good ventilation will reduce the dust hazard.

15.3. TOXIC MATERIALS

Hazard to workers

Some glaze materials are directly poisonous if eaten or inhaled. The effect is not immediate but accumulates in the body over the years. The most dangerous are raw lead materials. Lead compounds should only be used as a frit. Other toxic materials are:

Antimony oxide
Barium carbonate
Cadmium compounds (in color pigment)
Chromium dioxide
Cobalt oxide and carbonate
Copper oxide and carbonate
Nickel oxide
Zinc oxide

Preventive rules are:

- Wear a dust mask when dry mixing the materials.
- Wash hands after working with these materials.
- Wear special clothing only for working.
- Never eat, drink or smoke in the workshop.

Hazard to crockery users

The main danger for users of crockery is the release of lead from glazes. This may happend if the glaze contains free lead and the glaze is used for storing acidic food. Glazes made with leadfrits may be perfectly safe, but it depends very much on the composition of the glaze. Unless your crockery can be checked regularly by a chemical laboratory, it is safer not to use lead glazes for items meant for food.

16. GLAZE FORMULA CALCULATIONS

Glazes are expressed in several different forms:

Recipe: a list of *actual materials* and *weights,* used directly to make the glaze.

Molecular formula: shows the *relative proportion* of *molecules* of flux, alumina and silica in the glaze. Must be converted to recipe to make the glaze.

Chemical analysis: shows the percentage of oxides in the glaze. Also known as *ultimate composition.*

Seger formula: a special molecular formula, which makes it easier to compare glazes. It is also known as the "empirical formula".

FIGURE 16.0.0.A The tedious job of ceramics calculation is well suited for computers.

16.1. GLAZE FORMULA CHEMISTRY

Why glaze formulas?

As you already know, glaze materials are complicated and if you only work with clay, limestone, talc, quartz etc. there is no way to theoretically understand how they combine in the glaze. For this reason, in order to make glazing scientific and systematic, it is necessary to use chemistry. This makes it possible to write materials as chemical symbols and to make calculations that help to invent new glazes and to alter existing recipes.

16.1.1. USING CHEMICAL SYMBOLS

Chemical symbols are a language for describing atoms, molecules and the way they are combined to make up the various materials used in chemistry and in glazes.

As already described at the beginning of the book, there are more than 100 *elements,* which are the basic building blocks of glaze materials. Each one has a chemical symbol:

Calcium = Ca
Copper = Cu
Iron = Fe
etc.

16.1.2. CHEMICAL REACTIONS

Elements are usually not found by themselves in nature. The basic nature of elements is to combine with each other: this process is called a *chemical reaction* and takes place in nature through the effects of heat, pressure etc. When elements combine, they are called compounds and they can be described by chemical formulas, which show the number of atoms and how they are attached to each other.

For example, china clay is written as $Al_2O_3 \cdot 2SiO_2 \cdot 2H_2O$. Each element is followed by a number written below the line: this is the number of atoms in the compound.

Al_2 means 2 atoms of alumina and O_3 means 3 atoms of oxygen. This is the compound *aluminum oxide.*

If no number follows the element symbol, it is understood to be only 1 atom.
The raised period (·) shows that the compounds are joined together chemically to form a complex compound. The large numbers before each compound mean the number of molecules that combine. If there is no number in front, it is understood to mean 1 molecule.

Al_2O_3 means 1 molecule of aluminum oxide.

$2SiO_2$ means 2 molecules of silicon oxide.

So $Al_2O_3 \cdot 2SiO_2 \cdot 2H_2O$ is a complex compound consisting of 1 molecule of aluminum oxide, 2 molecules of silicon oxide and 2 molecules of water.

These compounds cannot be broken down physically but can combine with other compounds when heated sufficiently in the kiln.

16.1.3. MOLECULAR WEIGHTS

Each kind of molecule has a specific weight. We all know that 1 kg of lead is much smaller than 1 kg of aluminum. This is because the molecules are heavier and are packed together more closely.

Because it is impossible to weigh individual molecules, they have all been assigned *molecular weights,* which are relative to hydrogen, which has been given the molecular weight of 1. The molecular weights of all the other elements are based on how much heavier they are compared to hydrogen.

So the molecular weight of oxygen = 16, meaning it is 16 times heavier than hydrogen.

16.1.4. FORMULA WEIGHT OF MINERALS

The molecular weights of all the elements in a compound can be added together to get the total molecular weight. This is called the *formula weight.* In our example of kaolin clay, we can look in the table of elements and oxides in the appendix to find out the individual molecular weights. Molecular weight is abbreviated to "MW". In order to simplify calculations we round up the MW figures. This is accurate enough since we seldom know the exact composition of our raw materials anyway.

$Al_2O_3 \cdot 2SiO_2 \cdot 2H_2O$

ELEMENT	MW	NUMBER OF ATOMS		OXIDE WEIGHT	COMPOUND WEIGHT
Al	27	2	2 X 27 = 54		
O	16	3	3 X 16 = 48	102	1 X 102 = 102
Si	28	1	1 X 28 = 28		
O	16	2	2 X 16 = 32	60	2 X 60 = 120
H	1	2	2 X 1 = 2		
O	16	1	1 X 16 = 16	18	2 X 18 = 36

So the total MW of $Al_2O_3 \cdot 2SiO_2 \cdot 2H_2O$ = 258

This is known as *formula weight.*

16.1.5. PERCENTAGE TO FORMULA

Glaze formulas are often given as percentages of the various oxides. In order to find out the chemical formula, the rule is to:

Divide each oxide by its molecular weight!

In the appendix you will find the molecular weight of glaze oxide and materials.

Example:
Calculation of molecular formula of kaolin with the following chemical composition:

OXIDE	SYMBOL	PERCENT	MW	CALCULATION
Silica	SiO_2	46.51%	60	46.51/60 = 0.775
Alumina	Al_2O_3	39.53%	102	39.53/102 = 0.387
Water	H_2O	13.96%	18	13.96/18 = 0.775

The molecular formula is $0.387Al_2O_3 \cdot 0.775\ SiO_2 \cdot 0.775H_2O$

Because this is difficult to use, we divide all the numbers by the smallest one.

0.387/0.387 = 1 Al_2O_3
0.775/0.387 = 2 SiO_2
0.775/0.387 = 2 H_2O

The formula comes out neatly as the familiar $Al_2O_3 \cdot 2SiO_2 \cdot 2H_2O$, or kaolin!

For using a material in glaze calculation we need to calculate its formula weight. This is done as shown above for kaolin.

16.2. SEGER FORMULA

About 100 years ago a German ceramist, Hermann Seger, developed Seger cones for measuring temperatures in kilns. He also proposed writing the composition of glazes according to the number of different oxides in the glaze instead of listing the raw materials used in the glaze.

For example: Aluminum oxide can be added to the glaze either in the form of clay ($Al_2O_3 \cdot 2SiO_2 \cdot 2H_2O$) or feldspar ($K_2O \cdot Al_2O_3 . 6SiO_2$).

The oxides used in glazes are divided into three groups according to the way the oxides work in the glaze.

Fluxes
This group of oxides functions as melter, and fluxes are also called basic oxides or bases. They are written RO or R_2O, where R represents any atom and O represents oxygen. So all the fluxes are a combination of one or two element atoms and one oxygen atom.

Stabilizers
These work as stiffeners in the melted glaze to prevent it from running too much. They are considered neutral oxides and are written as R_2O_3 or two atoms of some element combined with three oxygen atoms.

Glass formers
These form the noncrystalline structure of the glaze. They are called acidic oxides and are written as RO_2 or one element atom combined with two oxygen atoms.

Seger formulas allow all glaze formulas to be expressed in a table, keeping the groups separate in order to make comparison of different formulas easy (see below).

In the table form, the sum of the fluxes must always equal 1, which makes different formulas easy to compare.

Examples
The organization of the Seger formula is always according to the table shown below.

FLUXES	STABILIZER	GLASS FORMERS
RO, R_2O	R_2O_3	RO_2
Alkalis:	Al_2O_3	SiO_2
K_2O	B_2O_3	TiO_2
Na_2O		B_2O_3
Li_2O		
Alkaline earths:		
CaO		
MgO		
BaO		
Other:		
PbO		
ZnO		

Note: B_2O_3 is sometimes listed under stabilizers and sometimes under glass formers, since it has both characteristics.

TABLE OF LIMIT FORMULAS*

NOTE: "KNaO" is a symbol for either sodium or potassium oxide.

c012 – 08 Lead Glazes

PbO	0.7 – 1.0	Al_2O_3	0.05 – 0.2	SiO_2	1.0 – 1.5
KNaO	0 – 0.3				
ZnO	0 – 0.1				
CaO	0 – 0.2				

c08 – 01 Lead Glazes

PbO	0.7 – 1.0	Al_2O_3	0.1 – 0.25	SiO2	1.5 – 2.0
KNaO	0 – 0.3				
ZnO	0 – 0.2				
CaO	0 – 0.3				

c08 – 04 Alkaline Glazes

PbO	0 – 0.5	Al_2O_3	0.5 – 0.25	SiO2	1.5 – 2.5
KNaO	0.4 – 0.8				
ZnO	0 – 0.2				
CaO	0 – 0.3				

c08 – 04 Lead-Boron

PbO	0.2 – 0.6	Al_2O_3	0.15 – 0.2	SiO_2	1.5 – 2.5
KNaO	0.1 – 0.25			B_2O_3	0.15 – 0.6
ZnO	0.1 – 0.25				
CaO	0.3 – 0.6				
BaO	0 – 0.15				

c2 – 5 Lead Glazes

PbO	0.4 – 0.6	Al_2O_3	0.2 – 0.28	SiO2	2.0 – 3.0
KNaO	0.1 – 0.25				
ZnO	0 – 0.25				
CaO	0.1 – 0.4				

c2 – 5 Boron

KNaO	0.1 – 0.25	Al_2O_3	0.2 – 0.28	SiO2	2.0 – 3.0
ZnO	0.1 – 0.25			B_2O_3	0.3 – 0.6
CaO	0.2 – 0.5				
BaO	0.1 – 0.25				

c2 – 5 Lead Borosilicate

PbO	0.2 – 0.3	Al_2O_3	0.25 – 0.35	SiO2	2.5 – 3.5
KNaO	0.2 – 0.3			B_2O_3	0.2 – 0.6
ZnO	0 – 0.1				
CaO	0.35 – 0.5				

c8 – 12 Stoneware and Porcelain

KNaO	0.2 – 0.4	Al_2O_3	0.3 – 0.5	SiO2	3.0 – 5.0
ZnO	0 – 0.3			B_2O_3	0.1 – 0.3
CaO	0.4 – 0.7				
BaO	0 – 0.3				
MgO	0 – 0.3				

* D. Rhodes: *Clay and Glazes for the Potter.*

For example, a simple unfritted lead glaze would look like this:

FLUXES	STABILIZER	GLASS FORMERS
RO, R_2O	R_2O_3	RO2
PbO 1.0	Al_2O_3 0.1	SiO_2 1.5

Remember that the flux column always totals 1.0.

A more complicated formula is the unfritted boron glaze:

CaO	.414	Al_2O_3	.322	SiO_2	2.291
MgO	.414			B_2O_3	.931
K_2O	.172				
	1.000				

There are some basic rules for the ratio of oxides in the 3 different groups, according to glaze temperature. These are called limit formulas (see page 139). They should only be considered guidelines, as many glazes exceed the limits in practice.

- Addition of 0.1 part SiO_2 to a glaze will increase the melting point by about 20°C.

- Addition of 0.05 part B_2O_3 will lower the melting point by 20°C.

The formulas of pyrometric Seger cones are listed in the appendix. These can also be used as a guide for glazes by choosing a cone formula 4 to 5 cones below the glaze firing temperature. If you need a glaze for cone 9, 1280°C, you can use the cone 5 formula for the glaze.

16.2.1. BENEFITS OF USING SEGER FORMULA

The main usefulness of the Seger formula is that it presents glazes in a way that is easy to compare. It is used for:

Originating new glazes
Glazes with desired characteristics of color, mattness etc. can first be written as Seger formulas, selecting oxides that are known to produce the effects.

Comparing glaze recipes
It is difficult to look at two recipes and see how they are different. If they are converted into Seger formulas, the differences can easily be seen.

Substituting materials
If a material is no longer available, other materials can be substituted by working out the quantities in the Seger formula.

Modifying glazes
Glazes that change character, have problems etc. can be analyzed as Seger formulas, and directions for testing decided.

The Seger formula should be considered a guide only, as most theoretical glazes do not react as expected and still require empirical testing to develop them fully. If you want to use Seger formulas for your glazes it is nice to have exact chemical analysis of your raw materials, but this is seldom the case. Instead you will have to pick one of the materials listed in the appendix. They may be close enough for practical work.

16.2.2. GLAZE RECIPE FROM FORMULA

To get the glaze recipe from the formula, there is a standard series of calculations.

Simple lead glaze example

PbO 1.0 Al_2O_3 0.1 SiO_2 1.5

First decide which raw materials to use. For lead oxide, PbO, the choices are red lead, white lead or litharge. Al_2O_3 is almost always obtained from china clay, and SiO_2 usually from quartz powder.

The calculation is helped by a table like this:

Material and formula	Mol. parts	PbO 1.0	Al_2O_3 0.1	SiO_2 1.5
Litharge, PbO	1.0	1.0		
Kaolin, $Al_2O_3 \cdot 2SiO_2 \cdot 2H_2O$	0.1		0.1	0.2
Quartz, SiO_2	1.3			1.3
Total		1.0	0.1	1.5

1.0 molecular part (MP) of litharge provides all PbO needed. We enter kaolin and its formula in the table and write 0.1 for MP. When we take 0.1 part kaolin, we get 0.1 Al_2O_3 and we enter this on the right. In the kaolin formula we have 2 SiO_2 so when we take 0.1 kaolin we get 0.2 SiO_2. We list this under SiO_2. We need 1.5 SiO_2 so 1.3 remains and we get this from quartz.

Next the required molecular parts, MP, of each material are multiplied by their molecular weights, MW, to get the batch weight of each material:

Material	MP	MW	Calculation	Batch weight
Litharge	1.0	223	223 X 1	223
Kaolin	0.1	258	258 X 0.1	25.8
Quartz	1.3	60	60 X 1.3	78.0

To change the recipe into percentages, all the figures are divided by the total:

Litharge 223/326.8 = .68 = 68%

Kaolin 25.8/326.8 = .08 = 8%

Quartz 78.0/326.8 = .24 = 24%

Boron glaze example

A more complicated formula is the unfritted boron glaze.

CaO	.414	Al_2O_3	.322	SiO_2	2.291
MgO	.414			B_2O_3	.931
K_2O	.172				

Again, the first step is to select materials. Because materials that supply more than one oxide usually work better in glazes, they are preferred if available. We need both CaO and MgO, which are supplied by dolomite, $CaCO_3 \cdot MgCO_3$. Potash feldspar supplies K_2O along with Al_2Ok_3 and SiO_2. Quartz provides SiO_2. For boron, boric acid is selected.

Use a calculation table and follow these steps:

CALCULATION PROCEDURE

1. Enter formula at top of calculation table.
2. Select materials, enter formula and MW.
3. Multiply each material's MW with its MW and enter result in part's weight.
4. Enter MP of each oxide of the material under the formula to check oxide balance.
5. Convert parts' weight into a percentage recipe.

Material and formula	MW	MP	MP's weight	CaO .414	MgO .414	K_2O .172	Al_2O_3 .322	B_2O_3 .931	SiO_2 2.291
Dolomite $CaCO_3 \cdot MgCO_3$	184	.414	76	.414	.414				
Feldspar $K_2O \cdot Al_2O_3 \cdot 6SiO_2$	556	.172	96			.172	.172		1.032
Kaolin $Al_2O_3 \cdot 2SiO_2 \cdot 2H_2O$	258	.15	39				.15		.3
Quartz SiO_2	60	.959	58						.959
Boric acid $B_2O_3 \cdot 3H_2O$	123	.931	115					.931	
Total			384	.414	.414	.172	.322	.931	2.291

As before we change the recipe to percentage:

Dolomite	$76/384 \times 100 = 19.8$	20%
Potash feldspar	$96/384 \times 100 = 25$	25%
Kaolin	$39/384 \times 100 = 10.2$	10%
Quartz	$58/384 \times 100 = 15.1$	15%
Boric acid	$115/384 \times 100 = 29.9$	30%

When calculating from formula to recipe, there is no need to carry out results beyond round figures, particularly when we do not know the exact chemical analysis of our materials.

FIGURE 16.2.2.A Copy this example of a calculation table.

Material and Formula	M.W.	M.P.	parts weight	Oxides in formula							

16.2.3. FORMULA FROM GLAZE RECIPE

Calculating from a recipe to the Seger formula is the same process in reverse. We will use the same raw boric acid glaze as an example. Again we use the calculation table and the following steps:

1. Enter recipe materials and their formulas in the left column and MW and recipe figures in MP's weight column.
2. Write oxides of the materials at top of table.
3. Divide each recipe figure with its MW and enter result under MP.
4. Multiply MP with each oxide in material formula and enter result under respective oxide in the right columns.
5. Add together all oxides and list them according to RO-R_2O_3-RO_2.
6. Add oxides in RO and divide all RO figures with the total.

Material and formula	MW	MP	MP's weight	CaO	MgO	K_2O	Al_2O_3	B_2O_3	SiO_2
Dolomite $CaCO_3 \cdot MgCO_3$	184	.109	20	.109	.109				
Feldspar $K_2O \cdot Al_2O_3 \cdot 6SiO_2$	556	.045	25			.045	.045		.270
Kaolin $Al_2O_3 \cdot 2SiO_2 \cdot 2H_2O$	258	.039	10				.039		.078
Quartz SiO_2	60	.250	15						.250
Boric acid $B_2O_3 \cdot 3H_2O$	123	.244	30					.244	
Total			384	.109	.109	.045	.084	.244	.598

Note that from dolomite only CaO and MgO are entered in the formula. CO_2 is released during heating and does not take part in the glaze melt. H_2O of kaolin and boric acid likewise evaporates.

The oxides are set up in the standard Seger formula:

K_2O	.045	Al_2O_3	.084	SiO_2	.598
CaO	.109			B_2O_3	.244
MgO	.109				
	.263				

The formula is brought to unity by dividing all the figures by the total, .263, in the left column.

K_2O	.171	Al_2O_3	.319	SiO_2	2.27
CaO	.414			B_2O_3	.928
MgO	.414				

NOTE: The figures are not exactly the same as the original formula above, due to rounding off the figures. This is accurate enough for practical work.

If you have a chemical analysis of materials you want to use in a glaze, you first have to calculate the formula of the material as described on page 137. Then you enter this formula and its formula weight in the table under MW.

16.3. FRIT CALCULATION

Frit calculation is done in the same way as calculating a glaze, but the calculation is slightly more complicated. As with glazes it is important to follow the recipes accurately.

This also means that you have to make sure that the raw materials are not wet when you weigh them. Also remember that materials like calcined soda and borax will absorb moisture from the air if they are not kept in a sealed container.

16.3.1. MOISTURE COMPENSATION

If you have to weigh materials with a high moisture content you can compensate for this. Weigh 100 g of the material, dry it and then weigh it again. Moisture content is:

$$\frac{\text{wet weight} - \text{dry weight} \times 100}{\text{dry weight}} = x\ \%$$

This x % is added to the amount you are weighing to compensate for its moisture content. Example:

100 g kaolin weighs 92 g after drying.

$$\frac{100 - 92}{92} \times 100 = 8.7\ \%$$

kaolin in recipe	3500 g
compensation 8.7 % × 3500 =	304.5 g
total amount needed	3804.5 g

16.3.2. FORMULA RULES FOR FRIT

The practice of fritting was described in section 7. The main reason for fritting is to make glaze materials insoluble, which is possible if the frit materials are mixed in the right proportion. In formula terms they should fall within these limits:

- Ratio flux : SiO_2 should be between 1:1.5 and 1:3.
- The sum of K_2O and Na_2O should not exceed 0.5 molecular parts on the flux side, the rest being other fluxes like PbO, CaO, ZnO, BaO.
- B_2O_3 to SiO_2 not less than 1:2, but with other materials like PbO, CaO, MgO, K_2O in the frit the proportion can go down to 1:1.5.
- A little Al_2O_3, at least 0.05 mol. parts, reduces solubility but it should not exceed 0.2 mol. parts because it reduces the fluidity of the frit melt.

16.3.3. FRIT BASED ON GLAZE FORMULA

We have a glaze formula of an opaque boron glaze for 1100°C:

K_2O	.23	Al_2O_3	.30	SiO_2	2.60
ZnO	.27			B_2O_3	.80
CaO	.50				

Initially we calculate the recipe as it was done for the unfritted glaze. We get the K_2O from potash feldspar. Borax cannot be used for boric oxide because no Na_2O is needed in the formula and so boric acid is requiered. We get the CaO from whiting and the rest of the materials will be kaolin, quartz and zinc oxide.

We now decide what material to include in the frit batch and what to include in the ball milling only. This is done according to the above rules. We need to include all the soluble boric acid. Along with that we can also include whiting and zinc oxide and some potash feldspar but not all because its Al_2O_3 will reduce the frit's fluidity.

A frit formula could be:

K_2O	.1	Al_2O_3	.1	SiO_2	1.60
ZnO	.27			B_2O_3	.80
CaO	.50				

In this case we do not convert flux side to 1 because we need to compare it with the original formula:

	K_2O	ZnO	CaO	Al_2O_3	B_2O_3	SiO_2
Original:	.23	.27	.50	.30	.80	2.60
Frit:	.1	.27	.50	.10	.80	1.60
To mill:	.13	–	–	.20	–	1.00

Frit formula:

Material and formula	MW	MP	MP's weight	K_2O .10	ZnO .27	CaO .50	Al_2O_3 .10	B_2O_3 .80	SiO_2 1.60
Feldspar $K_2O \cdot Al_2O_3 \cdot 6SiO_2$	556	.10	55.6	.10			.10		.60
Whiting $CaCO_3$	100	.50	50.0			.50			
Quartz SiO_2	60	1.00	60.0						1.00
Zinc oxide ZnO	81	.27	21.9		.27				
Boric acid $B_2O_3 \cdot 3H_2O$	123	.80	98.4					.80	
Total			285.9	.10	.27	.50	.10	.80	1.60

Formula for materials added to the frit in the ball mill:

Material and formula	MW	MP	MP's weight	K_2O .13	Al_2O_3 .20	SiO_2 1.00
Feldspar $K_2O \cdot Al_2O_3 \cdot 6SiO_2$	556	.13	72.3	.13	.13	.78
Kaolin $Al_2O_3 \cdot 2SiO_2 \cdot 2H_2O$	258	.07	18.1		.07	.14
Quartz SiO_2	60	.08	4.8			.08
Total			95.2	.13	.27	1.00

Raw Frit Recipe

Potash feldspar	55.6	19.4%
Whiting	50.0	17.5%
Quartz	60.0	21.0%
Zinc oxide	21.9	7.7%
Boric acid	98.4	34.4%

Glaze Recipe

Raw frit	285.9	75.0%
Potash feldspar	72.3	19.0%
Kaolin	18.1	4.7%
Quartz	4.8	1.3%

One problem still remains. When the frit melts, a large amount of H_2O and CO_2 is lost. This loss does not influence the recipe if we weigh the raw frit materials, melt the frit and use all the melted frit in the glaze, adding the other material according to the original amount of raw frit. But it is much more practical to produce a large batch of frit at a time and later weigh the melted frit to produce smaller batches of glaze. We need to find out how much weight is lost.

16.3.4. FRIT LOSS CALCULATION

Practical loss

The loss can be found simply by weighing the amount of melted frit that is produced from a batch of frit. Example:

Raw frit batch weighs in total 500 kg.
After firing the (dry) frit weighs 280 kg.

$$\text{Loss in \%} = \frac{500 - 280}{500} \times 100 = 44\ \%$$

Theoretical loss

The loss can also be calculated based on the formula of the frit. On heating, whiting changes to calcium oxide:

$CaCO_3$ + heat -> CaO + CO_2

Only CaO enters the melted frit and we can calculate how much this weighs:

The MW of calcium carbonate is 100 and that of calcium oxide is 56 so loss is 44 parts. In percentage this is 44 % The number used to find the amount of oxide entering fusion is called the *conversion factor*, CF. In the material table in the appendix one column lists the conversion factor for all materials. At the bottom of the left column there is a list for the most common frit materials.

Material	CF	% loss
Barium carbonate	0.777	22.3
Borax (crystal)	0.526	47.4
Boric acid	0.563	43.7
Dolomite	0.523	47.7
Kaolin	0.861	13.9
Lead carbonate (white)	0.863	13.7
Lead oxide (red)	0.977	2.3
Magnesium carbonate	0.478	52.2
Pearl ash	0.682	31.8
Soda ash	0.585	41.5
Soda crystals	0.217	78.3
Whiting	0.561	43.9

Frit glaze example

We can now calculate the loss of our frit from before.

Frit Recipe:	Raw	CF	Melted
Potash feldspar	55.6		55.6
Whiting	50.0	x 0.561	28.1
Quartz	60.0		60.0
Zinc oxide	21.9		21.9
Boric acid	98.4	x 0.563	55.4
Total	285.9		221.0

Theoretically we get only 77.3 % melted frit from our raw frit batch. We found that 286.3 parts raw frit equal 221.2 parts melted frit so finally we can establish our glaze recipe based on melted frit:

Final glaze recipe:

Frit	221.0	69.9%
Potash feldspar	72.3	22.9%
Kaolin	18.1	5.7%
Quartz	4.8	1.5%

16.3.5. GLAZE RECIPE WITH STANDARD FRIT

Very often a ceramics producer gets the frit from a commercial supplier or wants to use only a few standard frits. Above we calculated a new frit based on the glaze formula. We will now calculate a glaze recipe from formula using a standard frit instead.

Example of a standard frit formula:

K_2O	.26	Al_2O_3	.05	SiO_2	2.5
ZnO	.13			B_2O_3	1.0
CaO	.61				

We will try to use the frit for the following glaze:

K_2O	.30	Al_2O_3	.40	SiO_2	3.5
ZnO	.20			B_2O_3	0.8
CaO	.50				

The calculation is done as with the unfritted glaze. First oxides are entered at the top of the table and we start to select materials to satisfy them. Before starting, we need to know the formula weight of the frit. In the appendix we get the MW of all the oxides and these we total.

K_2O	.26 x 94	=	24.4
Na_2O	.13 x 62	=	8.1
CaO	.61 x 56	=	34.2
Al_2O_3	.05 x 102	=	5.1
SiO_2	2.5 x 60	=	150.0
B_2O_3	1.0 x 70	=	70.0
Frit MW			291.8
This we round off to			292

The frit is entered in the calculation table like other materials with many oxides. The MP is selected according to the need of B_2O_3. It takes 0.8 MP of frit to get the needed 0.8 B_2O_3 and all the oxides listed in the frit formula are multiplied by this number and the results entered on the right of the table.

Material and formula	MW	MP	MP's weight	K_2O .30	Na_2O .20	CaO .50	Al_2O_3 .40	B_2O_3 .80	SiO_2 3.50
Frit	292	.80	233.6	.208	.104	.49	.040	.80	2.000
Feldspar $K_2O \cdot Al_2O_3 \cdot 6SiO_2$	556	.092	51.2	.092			.092		.552
Feldspar $Na_2O \cdot Al_2O_3 \cdot 6SiO_2$	524	.096	50.3		.096		.096		.576
Kaolin $Al_2O_3 \cdot 2SiO_2 \cdot 2H_2O$	258	.172	42.7				.172		.372
Whiting $CaCO_3$	100	.01	1.0			.01			
Total			378.8	.300	.200	.500	.400	.80	3.500

Glaze Recipe	Parts	%
Frit	233.6	61.7
Potash feldspar	51.2	13.5
Soda feldspar	50.3	13.3
Kaolin	42.7	11.2
Whiting	1	0.3

16.3.6. HINTS FOR USING UNKNOWN LOCAL MATERIALS

We have already discussed above calculating local materials by guessing their closest theoretical formula. This will usually give a good starting point for making line blends, which then can be used to get a working glaze or frit.

What do you do when you have a recipe or formula but do not know the analysis of your local materials and cannot get pure ones? Usually you can create a glaze using the formula or recipe as a starting point, but it is unlikely to match the description in the book.

The most common local materials are usually:

Clays

Common clays can be used in most glazes instead of kaolin, since they all contain Al_2O_3 and SiO_2. But they will have lower melting points and probably change the glaze color, since they will introduce K_2O, Na_2O, Fe_2O_3, CaO, MgO and perhaps other fluxes. Probably the easiest way to work with them is simply to substitute directly for the kaolin, fire a sample and then use it as the basis for line blends to get a working glaze.

Feldspars

There are a tremendous number of different feldspars, all of which vary in the relative amounts of K_2O, Na_2O, CaO, MgO, Al_2O_3 and SiO_2 they supply. This means that directly substituting feldspars will affect the melting point of the glaze, and possibly its color response. Try them out as direct substitutions, and then the result can be altered using line blends. If the new glaze seems underfired (dry surface), the fluxes can be increased. If it seems overfired (too fluid), the clay content can be increased.

CaO sources

Calcium is introduced into glazes from a large variety of raw materials: calcium carbonate, whiting, limestone, marble, seashells, coral, agricultural lime, etc. Usually, substituting will not make much difference, but again the result can be developed using line blends of the new material.

Glass cullet

Glass cullet means waste glass, which can be used as the basis for cheap glazes. The best glass to use is window glass, which can usually be obtained free of charge or cheap from glass suppliers. Window glass consists of soda-lime-silica and can be used as a frit in glazes. It melts at about 1100°C. With the addition of some flux and clay, it can be made into a low temperature glaze. However, because of its high CE, it will usually craze.

Unknown materials

If you find new materials that are completely unknown, the easiest way to find out what they do is to first fire a small sample of the material alone, to see if it melts or not and what color it becomes. If it melts, it is a strong flux. If it does not melt, it may still be a flux. Check the test carefully to see if it has reacted with the clay body. If it develops a strong color, it will probably affect the glaze colour.

The material should also be tested by adding it to a known glaze recipe as a line blend.

Appendix

Glaze Recipes

Glaze recipes are included in this book with a word of caution: Because of the wide variations in raw materials around the world, the same recipe will produce different results in different locations.

So please consider these recipes to be good starting points. They are not guaranteed to work without some modifications but will put you in the general area of success. Most of the glazes are not described as glossy, matt, opaque etc. Try them out and modify them according to what you have learned in this book.

With regard to frits these vary from one manufacturer to another. As frit making is not economical for the small potter, it is suggested to substitute locally available frits. The recipes are compiled from many different sources, which are listed below.

Fritted boron glazes for low temperatures

GLAZE #1.[1]

Temperature: 980 °C

Frit recipe

Borax	15.5
Potash feldspar	34.8
Marble	17.9
Boric acid	31.8

Glaze recipe

Frit	35.7
Potash feldspar	25.8
Kaolin	1.0
Quartz	17.2
Soda feldspar	20.3

Glaze formula

0.25	K_2O	0.40	Al_2O_3	3.50	SiO_2
0.15	Na_2O			1.00	B_2O_3
0.60	CaO				

GLAZE #2.[1]

Temperature: 980°C

Frit recipe

Borax	33.3
Potash feldspar	21.1
Marble	16.6
Kaolin	9.0
Quartz	20.0

Glaze recipe

Frit	35.1
Potash feldspar	24.3
Magnesium carbonate	4.3
Barium carbonate	8.2
Kaolin	11.9
Quartz	17.2

Glaze formula

0.17	K_2O	0.47	Al_2O_3	3.42	SiO_2
0.16	Na_2O			0.83	B_2O_3
0.25	CaO				
0.25	MgO				
0.17	BaO				

GLAZE #3.[1]

Temperature: 1080 °C

Glaze recipe

Frit from glaze #2	35.9
Potash feldspar	16.7
Kaolin	15.5
Quartz	14.4
Marble	12.0
Dolomite	5.5

Glaze formula

0.40	K_2O	0.30	Al_2O_3	2.80	SiO_2
0.50	CaO			0.60	B_2O_3
0.10	MgO				

GLAZE #4.[2]

Temperature: 1100 °C

Frit recipe

Borax	10.0
Potash feldspar	33.5
Marble	15.3
Kaolin	4.0
Quartz	12.8
Boric acid	24.4

Glaze recipe

Frit	94.0
Kaolin	6.0

Glaze formula

0.34	K_2O	0.34	Al_2O_3	3.5	SiO_2
0.66	CaO			1.0	B_2O_3

GLAZE #5.[3]

Temperature: 1040 °C

Clear transparent glaze, will craze on most bodies.

Frit recipe

Borax	47.50
Boric acid	10.50
Potash feldspar	10.50
Quartz	21.00
Kaolin	8.50
Zinc oxide	2.00

Glaze recipe

Frit	62.00
Local red clay	12.00
Kaolin	8.00
Quartz	8.00
Potash feldspar	5.00
Barium carbonate	5.00

Glaze formula

0.11	K_2O	0.48	Al_2O_3	3.83	SiO_2
0.53	Na_2O				
0.16	BaO			1.36	B_2O_3
0.10	MgO				
0.10	ZnO				

Calculation of this formula is based on chemical analysis of local glaze materials.

GLAZE #6.[3]

Temperature: 1040 °C

Semiopaque glaze.

Frit recipe

Borax	43.70
Marble	2.40
Talc	0.60
Zinc oxide	0.30
Kaolin	0.60
Rice husk ash*	40.00
Quartz	7.00

* About 30 parts quartz can replace the rice husk ash.

Glaze recipe

Frit	37.7
Wollastonite	5.0
Kaolin	16.0
Potash feldspar	6.0
Quartz	7.8
Zinc oxide	7.5
Zircon	20.0

Glaze formula

0.04	K_2O	0.37	Al_2O_3	4.37	SiO_2
0.27	Na_2O			0.16	ZrO_2
0.39	ZnO			0.48	B_2O_3
0.30	CaO				

Calculation of this formula is based on chemical analysis of the local glaze materials. The content of zirconium oxide in the zircon used in this recipe is only half of what is normal for standard quality zircon.

GLAZE #7.[3]

Temperature: 1040 °C
Opaque, silky matt glaze.

Frit recipe

Borax	40.5
Marble	2.2
Talc	0.6
Zinc oxide	0.2
Kaolin	5.5
Quartz	43.7
Bone ash calcined	7.3

Glaze recipe

Frit	57.0
Kaolin	9.0
Ball clay	8.0
Quartz	11.0
Marble	2.0
Zinc oxide	13.0

Glaze formula

0.05	MgO	0.26	Al_2O_3	2.90	SiO_2
0.25	Na_2O				
0.53	ZnO			0.50	B_2O_3
0.17	CaO				

Nonfritted borax glazes

GLAZE #8.[3]

Temperature: 1040 – 1080 °C

Glaze recipe

Borax	7.2
Black local clay	15.00
Glass cullet	48.00
Barium carbonate	15.00
Manganese dioxide	15.00

Black glossy glaze used for tiles and sewage pipes.

Glaze formula

0.25	Na_2O	0.06	Al_2O_3	1.25	SiO_2
0.17	BaO				
0.16	CaO			0.09	B_2O_3
0.11	MgO				
0.31	MnO				

GLAZE #9.[3]

Temperature: 1020 – 1040 °C

Glaze recipe

Borax	50.00
Local red clay	5.00
Rice husk ash	30.00
Whiting	4.00
Yellow ochre	11.00

Transparent, glossy glaze.

Glaze formula

0.73	K_2O	0.11	Al_2O_3	1.71	SiO_2
0.17	CaO				
0.10	MgO			1.34	B_2O_3

GLAZE #10.[4]

Temperature: 1050 – 1100 °C

Glaze recipe

Borax	30.00
Potash feldspar	25.00
Quartz	15.00
Dolomite	20.00
Ball clay	10.00

Glaze formula

0.41	CaO	0.29	Al_2O_3	2.29	SiO_2
0.41	MgO				
0.17	K_2O			0.91	B_2O_3

GLAZE #11.[5]

Temperature: 1260 °C

Glaze recipe

Potash feldspar	16.6
Kaolin	11.0
Quartz	23.7
Marble	15.0
Borax	22.2
Soda ash	11.5

Glaze formula

0.09	K_2O	0.21	Al_2O_3	1.90	SiO_2
0.48	Na_2O				
0.43	CaO			0.36	B_2O_3

Glazes for heavy clay products

GLAZE #12.[6]

Temperature: 960 – 980 °C

Glaze recipe

Sodium silicate	34.0
Glass cullet	11.0
Zircon	30.0
Quartz	10.0
Kaolin	5.0

GLAZE #13.[7]

Temperature: 920 – 960 °C

Glaze recipe

Glass cullet	70.0
Local red clay	15.0
Whiting	15.0

Opaque low-cost glaze.

Glaze formula

0.56	CaO	0.13	Al_2O_3	1.94	SiO_2
0.32	Na_2O				
0.12	Mg				

GLAZE #14.[6]

Temperature: 1000 °C

Frit recipe

Red lead	30.00
Quartz	26.30
Potash feldspar	12.20
Whiting	7.00
Borax	10.50
Magnesium carbonate	7.00
Zinc oxide	7.00

Glaze recipe

Frit	92
Kaolin	8

Glaze formula

0.31	PbO	0.12	Al_2O_3	1.50	SiO_2
0.17	CaO				
0.20	MgO			0.13	B_2O_3
0.05	K_2O				
0.07	Na_2O				
0.20	ZnO				

GLAZE #15.[6]

Temperature: 1000 °C

Frit recipe

Red lead	17.80
Quartz	26.10
Potash feldspar	22.40
Whiting	10.30
Boric acid	16.40
Soda ash	3.00
Kaolin	4.00

Glaze recipe

Frit	92.0
Kaolin	8.0

Glaze formula

0.31	PbO	0.34	Al_2O_3	3.07	SiO_2
0.41	CaO				
0.16	K_2O			0.53	B_2O_3
0.11	Na_2O				

GLAZE #16.[8]

Temperature: 1050 °C

Frit recipe

Quartz	20.00
Potash feldspar	25.00
Whiting	7.00
Borax	25.00
Magnesium carbonate	5.00
Zinc oxide	8.00
Zircon	10.00

Glaze recipe

Frit	92.0
Kaolin	8.0

Glaze formula

0.21	CaO	0.22	Al_2O_3	2.11	SiO_2
0.18	MgO			0.16	ZrO_2
0.13	K_2O			0.39	B_2O_3
0.19	Na_2O				
0.29	ZnO				

GLAZE #17.[9]

Temperature: 950 – 1050 °C

Frit recipe

Quartz	28.30
Potash feldspar	40.80
Whiting	7.00
Borax	13.00
Soda ash	5.10
Titanium dioxide	5.60

Glaze recipe

Frit	94.0
Kaolin	6.0

Glaze formula

0.32	CaO	0.42	Al_2O_3	4.20	SiO_2
0.32	K_2O			0.31	TiO_2
0.36	Na_2O			0.34	B_2O_3

GLAZE #18.[10]

Temperature: 1000 – 1100 °C

Frit recipe

Red lead	30.56
Quartz	20.83
Whiting	18.05
Borax	25.00
Kaolin	5.56

Glaze recipe

Frit	94.0
Kaolin	6.0

Glaze formula

0.47	CaO	0.11	Al_2O_3	1.13	SiO_2
0.35	PbO				
0.17	Na_2O			0.34	B_2O_3

GLAZE #19.[11]

Temperature: 1050 – 1130 °C

Frit recipe

Quartz	34.20
Whiting	5.90
Borax	30.10
Zinc oxide	1.20
Boric acid	4.00
Kaolin	11.00
Zircon	12.70
Barium carbonate	1.00

Glaze recipe

Frit	95.0
Kaolin	3.0
Bentonite	2.0

Glaze formula

0.03	BaO	0.34	Al_2O_3	4.78	SiO_2
0.37	CaO			0.43	ZrO_2
0.50	Na_2O			1.19	B_2O_3
0.10	ZnO				

Stoneware glazes

GLAZE #20.[12]

Temperature: 1200 – 1250 °C

Glaze recipe

Potash feldspar	72.0
Kaolin	7.0
Quartz	8.0
Whiting	13.0

Glaze formula

0.50	CaO	0.60	Al_2O_3	3.72	SiO_2
0.50	K_2O				

GLAZE #21.[13]

Temperature: 1200 – 1250 °C

Glaze recipe

Local feldspar	62.0
Kaolin	5.5
Quartz	10.0
Whiting	11.0
Zinc oxide	5.0
Zircon	6.5

Glaze formula

0.274	K_2O	0.485	Al_2O_3	3.147	SiO_2
0.124	Na_2O			0.084	ZrO_2
0.286	CaO				
0.216	ZnO				

Calculation based on analysis of local materials.

GLAZE #22.[14]

Temperature: 1200 – 1250 °C

Glaze recipe

Local feldspar	53.0
Kaolin	2.0
Quartz	28.0
Whiting	17.0
Zircon	+ 10.0

Glaze formula

0.250	K_2O	0.403	Al_2O_3	3.912	SiO_2
0.113	Na_2O				
0.637	CaO				

GLAZE #23.[15]

Temperature: 1250 °C

Glaze recipe

Feldspar	40
Quartz	30
Whiting	20
Kaolin	10

Glaze formula

0.74	CaO	0.41	Al_2O_3	3.71	SiO_2
0.26	K_2O				

GLAZE #24.[16]

Temperature: 1180 – 1200 °C

Glaze recipe

Potash feldspar	46.5
Quartz	10.7
Barium carbonate	16.5
Colemanite	1.6
Zinc oxide	6.7

Glaze formula

0.24	BaO	0.24	Al_2O_3	1.97	SiO_2
0.28	CaO				
0.24	K_2O			0.41	B_2O_3
0.24	ZnO				

GLAZE #25.[17]

Temperature: 1200 °C

Glaze recipe

Feldspar	51.1
Quartz	5.9
Kaolin	10.8
Whiting	18.6
Zinc oxide	8.7
Bentonite	4.9

Glaze formula

0.48	CaO	0.37	Al_2O_3	2.03	SiO_2
0.17	K_2O				
0.05	Na_2O				
0.02	MgO				
0.27	ZnO				

GLAZE #26.[16]

Temperature: 1200 – 1250 °C

Glaze recipe

Dolomite	3.9
Kaolin	7.7
Nepheline syenite	53.8
Quartz	23.1
Whiting	11.5

Glaze formula

0.51	CaO	0.56	Al_2O_3	3.56	SiO_2
0.08	MgO				
0.10	K_2O				
0.31	Na_2O				

GLAZE #27.[16]

Temperature: 1250 °C

Glaze recipe

Granite	63.6
Dolomite	27.3
Kaolin	9.1

Glaze formula

0.41	CaO	0.31	Al_2O_3	1.99	SiO_2
0.39	MgO				
0.10	K_2O				
0.10	Na_2O				

GLAZE #28.[17]

Temperature: 1200 °C

Glaze recipe

Wood ash	50
Kaolin	5
Dolomite	3
Whiting	10
Bentonite	2
Feldspar	20
Quartz	5
Nepheline syenite	5

Glaze formula

0.51 CaO	0.09 Al_2O_3	0.73 SiO_2
0.22 MgO		
0.15 K_2O		
0.12 Na_2O		

GLAZE #29.[15]

Temperature: 1240 – 1300 °C

Glaze recipe

Feldspar	40
Ash	40
Pike clay	20

Glaze formula

0.35 CaO	0.17 Al_2O_3	1.27 SiO_2
0.25 MgO		
0.27 K_2O		
0.13 Na_2O		

Calculation based on oak ash.

Sources of recipes

Many of the glaze recipes are taken from *Ceramic Glazes*, Stefanov/Batschwarov, Bauverlag GmbH, Wiesbaden and Berlin, 1988, and the original sources are also mentioned.

1 D. Green: A Handbook of Pottery Glazes, Faber and Faber. London-Boston, 1978.
2 A. Blakely: Journal Amer. Ceram. Soc. 21, 1938, No. 7.
3 Ceramics Promotion Project, P.O. Box 2008, Kathmandu, Nepal.
4 Ali Sheriff, P.O. Box 175, Arusha, Tanzania.
5 A. Augustinik: Keramik, Stroiizdat, Moscow, 1975.
6 Z. Filipova: Steklo i Keramika 10, 1953, No. 3.
7 G. Alexeev: Steklo i Keramika 12, 1955, No. 9.
8 D. Eschenko: Herstellung und Anwendung von Fassadenkeramik, Stroiizdat, Moscow, 1967.
9 S. Batschwarov, P. Welikova: Patent Bulgaria No. 29923; 1980.
10 E. Rochwarger: Handbuch "Bauerkeramik", Stroiizdat, Moscow, 1976.
11 H. Hecht: Lehrbuch der Keramik, Urban and Schwarzenberg, Berlin, 1930.
12 D. Bilington: The Technique of Pottery, Batsford Ltd., London, 1962.
13 H. Norsker: Recipe used at a stoneware factory, Insein, Burma.
14 U. Kyin: Recipe used at a sanitary ware factory, Rangoon, Burma.
15 B. Leach: A Potter's Book, Faber and Faber, London, 1976.
16 W. Matthes: Keramische Glasuren, Verlag Müller, Darmstadt, 1985.
17 J. Chappel: The Potter's Complete Book of Clays and Glazes, Pitman. London, 1977.

Color Pigments

Below are listed some recipes for color pigments. In the chapter on color pigments you will find instructions on how to prepare them. These recipes will seldom work right away, but they can be used as starting points for developing color pigments based on local materials. As with all ceramic colors, the color depends very much on firing conditions, purity of the raw materials and composition of clay and glazes.

Cobalt Blue Pigments

	Sky blue[1]	Light blue[1]	Royal blue[1]	Green-blue[1]	Dark blue[1]	Dark blue[2]
Cobalt oxide	5	10	20	10	45	44.6
Alumina	90	60	60	10		55.4
Zinc oxide	5	30	20	80		
Kaolin					55	

Green Pigments

	Victoria green[2]	Bluish green[2]	Russian green[2]	Green[2]	Olive green[3]	Olive green[2]
Chrome oxide	25.0	25.8	15.0	30	19.6	32
Cobalt oxide		35.5	10.0			
Nickel oxide					8.7	16
Alumina		38.7				
Whiting	50.0					
Quartz	25.0		60.0	20	39.1	28
Feldspar			15.0	50		
Calcined borax					32.6	24

Black Pigments

	Black[4]	Black[4]	Black[3]	Dark brown[3]	Greenish black[2]	Brownish black[2]
Chrome oxide	44	17	32.3		65.5	48.8
Cobalt oxide	22	20	20.6			
Iron oxide	24	35	41.2	45.5	34.5	51.2
Manganese oxide	10	20	5.9	54.5		
Nickel oxide		8				

Pink/Yellow Pigments

	Pink[4]	Dark pink[1]	Deep yellow[1]	Yellow[3]	Titan yellow[2]	Naples yellow[2]
Antimony oxide			40			40
Chrome oxide	1	1.7				
Iron oxide			8			(+5)
Tin oxide	45					20
Quartz	30	54.5				
Whiting	24					
Red lead		3.5	40			40
Zinc oxide		40.3			50	
Soda ash			12			
Kaolin				11.3		
Feldspar				20.8		
Rutile				67.9	50	

Red/Brown Pigments

	Orange[3]	Red-brown[4]	Light red-brown[4]	Yellow-brown[4]	Chocolate brown[3]	Iron red[4]
Rutile	37.5					
Iron oxide		22.8	17.8	13.7	27.3	50.0
Chrome oxide		21.7	16.9	13.1	18.2	
Zinc oxide	16.7	55.5	53.9	55.6	50.0	
Alumina			11.4	17.6		50.0
Kaolin	16.7				4.5	
Tin oxide	29.1					

Violet/Turquoise

	Deep violet[5]	Dark violet[5]	Blue violet[5]	Turquoise[5]	Turquoise blue[5]
Manganese oxide	40	70	45		
Cobalt oxide	7		10	27	
Cobalt carbonate					2
Chrome oxide				18	
Copper oxide					40
Alumina				55	
Quartz	53		25		
Tin oxide		15			75
Kaolin		25			
Zinc oxide			20		

[1] S.K. Mirmira: Indian Pottery, CAPART, New Delhi, 1987.
[2] F. Singer & S. S. Singer: Industrial Ceramics. Chapman and Hall. London, 1963.
[3] C. F. Binns: The Manual of Practical Potting. Scott, Greenwood & Son. London, 1922.
[4] W. Lehnhäuser: Glasuren und ihre Farben. Wilhelm Knapp Verlag. Düsseldorf, 1959.
[5] J. L. Artigas: Formulario y Practicas de Ceramica. Gustavo Gili. Barcelona, 1961.

Ceramics Elements and Oxides

Element	Symbol	Atomic weight	Oxide	Formula	Molekular weight	
Aluminium	Al	27.1	Aluminium oxide	Al_2O_3	101.9	Alumina
Antimony	Sb	121.8	Antimony trioxide	Sb_2O_3	291.6	
Barium	Ba	137.3	Barium oxide	BaO	153.3	Baria
Boron	B	10.8	Boric oxide	B_2O_3	69.6	
Calcium	Ca	40.1	Calcium oxide	CaO	56.1	Calcia
Carbon	C	12.0	Carbon dioxide	CO_2	44.0	
Chromium	Cr	52.0	Chromium oxide	Cr_2O_3	152.0	
Cobalt	Co	58.9	Cobalt oxide	CoO	74.9	
Copper	Cu	63.5	Cupric oxide	CuO	79.5	
Hydrogen	H	1.0				
Iron	Fe	55.8	Ferric oxide	Fe_2O_3	159.6	Iron oxide
Lead	Pb	207.2	Lead monooxide	PbO	223.2	Litharge
Lithium	Li	6.9	Lithium oxide	Li_2O	29.8	Lithia
Magnesium	Mg	24.3	Magnesium oxide	MgO	40.3	Magnesia
Manganese	Mn	54.3	Manganese dioxide	MnO_2	86.9	
Nickel	Ni	58.7	Nickel oxide	NiO	74.7	
Oxygen	O	16.0				
Phosphorus	P	31.0	Phosphorus oxide	P_2O_5	142.0	
Potassium	K	39.1	Potassium oxide	K_2O	94.2	Potash
Silicon	Si	28.1	Silicon dioxide	SiO_2	60.1	Silica
Sodium	Na	23.0	Sodium oxide	Na_2O	62.0	Soda
Strontium	Sr	87.6	Strontium oxide	SrO	103.6	
Sulphur	S	32.1	Sulphur dioxide	SO_2	64.1	
Tin	Sn	118.7	Tin oxide	SnO_2	150.7	
Titanium	Ti	47.8	Titanium dioxide	TiO_2	79.9	Titania
Vanadium	V	50.9	Vanadium oxide	V_2O_5	181.8	
Zinc	Zn	65.4	Zinc oxide	ZnO	81.4	
Zirconium	Zr	91.2	Zirconium oxide	ZrO_2	123.2	Zirconia

The molecular weights are listed with one decimal point. For glaze formula calculations you can use round up figures, eg. for iron with MW 55.8 you round it up to 56.

Common Glaze Raw Materials

CF = conversion factor
MP = melting point, °C (degrees Celsius)
* = decomposing temperatur

NOTE: melting point and decomposing temperature are only relative indicators of how materials behave in glaze. Melting points are affected by combinations of materials! (See eutectics).

Material	Formula	Molecular Weight	Oxides after firing	Oxide Weight	CF	MP (°C)
Alumina	Al_2O_3	101.9	Al_2O_3	101.9	1	2050
Alumina Hydrate	$Al_2(OH)_6$	156	Al_2O_3	101.9	0.653	300*
Antimony oxide	Sb_2O_3	291.6	Sb_2O_3	291.6	1	652
Barium carbonate	$BaCO_3$	197.3	BaO	153.3	0.777	1450*
Barim sulfate (barytes)	$BaSO_4$	233.4	BaO	153.4	0.657	1350*
Bentonite	$Al_2O_3 \cdot 4SiO_2 \cdot H_2O$	360.4	Al_2O_3 SiO_2	101.9 60.1	0.283 0.667	600*
Bone ash	see calcium phosphate					
Borax	$Na_2B_4O_7 \cdot 1OH_2$	381.5	B_2O_3 Na_2O	69.6 62	0.364 0.162	200*
Borax, anhydrous	$Na_2B_4O_7$	201.3	Na_2O B_2O_3	62 69.6	0.308 0.692	237
Boric acid	$B_2O_3 \cdot 3H_2O$	123.7	B_2O_3	69.6	0.563	185*
Calcite	$CaCO_3$	100.1	CaO	56.1	0.561	620*
Calcium carbonate	$CaCO_3$	100.1	CaO	56.1	0.561	850*
Calcium phosphate	$Ca_3(PO_4)_2$	310.2	CaO P_2O_5	56.1 142.0	0.541 0.459	1550
Chrome oxide	Cr_2O_3	152.0	Cr_2O_3	152.0	1	2435
Cobalt carbonate	$CoCO_3$	119.0	CoO	74.9	0.629	800*
Cobalt oxide (black)	Co_3O_4	240.8	CoO	74.9	0.933	905*
Colemanite	$2CaO \cdot 3B_2O_3 \cdot 5H_2O$	412.0	CaO B_2O_3	56.1 69.6	0.272 0.508	1100
Copper carbonate	$CuCO_3 \cdot Cu(OH)_2$	221.2	CuO	79.5	0.720	200*
Copper oxide black	CuO	79.5	CuO	79.5	1	1148
Copper oxide red	Cu_2O	143.0	CuO	79.5	1.112	1230
Cristobalite	SiO_2	60.1	SiO_2	60.1	1	1713
Dolomite	$CaCO_3 \cdot MgCO_3$	184.4	CaO MgO	56.1 40.3	0.304 0.219	600*
Enstatite	$MgO \cdot SiO_2$	100.4	MgO SiO_2	40.3 60.1	0.401 0.599	1550
Epsom salt	$MgO \cdot SO_3 \cdot 7H_2O$	246.5	MgO	40.3	0.162	1120

Material	Formula	Molecular Weight	Oxides after firing	Oxide Weight	CF	MP (^oC)
Feldspar, potash	$K_2O \cdot Al_2O_3 \cdot 6SiO_2$	556.5	K_2O Al_2O_3 SiO_2	94.2 101.9 60.1	0.169 0.183 0.648	1200
Feldspar, soda	$Na_2O \cdot Al_2O_3 \cdot 6SiO_2$	524.3	Na_2O Al_2O_3 SiO_2	62.0 101.9 60.1	0.118 0.194 0.688	1200
Feldspar, lime	$CaO \cdot Al_2O_3 \cdot 2SiO_2$	278.1	CaO Al_2O_3 SiO_2	56.1 101.9 60.1	0.202 0.366 0.432	1550
Flint	SiO_2	60.1	SiO_2	60.1	1	1600
Glass (bottle)	CaO .46 MgO .02 Al_2O_3 .035 Na_2O .51 K_2O .01 SiO_2 2.9	237.0	SiO_2 Al_2O_3 CaO MgO Na_2O K_2O	60.1 101.9 56.1 40.3 62.0 94.2	0.735 0.015 0.109 0.003 0.134 0.004	1150
Gypsum	$CaSO_4 \cdot 2H_2O$	172.2	CaO	56.1	0.326	1360
Iron oxide (black)	FeO	71.8	Fe_2O_3	159.6	1.111	1420
Iron oxide (red)	Fe_2O_3	159.6	Fe_2O_3	159.6	1	1565*
Kaolin (China clay)	$Al_2O_3 \cdot 2SiO_2 \cdot 2H_2O$	258.1	Al_2O_3 SiO_2	101.9 60.1	0.395 0.466	610*
Lead bisilicat	$PbO \cdot 2SiO_2$	343.3	PbO SiO_2	223.2 60.1	0.650 0.350	815
Lead carbonate (white)	$2PbCO_3 \cdot Pb(OH)_2$	775.7	PbO	223.2	0.863	400*
Lead oxide (Litharge)	PbO	223.2	PbO	223.2	1	160*
Lead oxide (red)	Pb_3O_4	685.6	PbO	223.2	0.977	500*
Limestone	(see calcium carbonate)					
Lithium carbonate	Li_2CO_3	73.9	Li_2O	29.9	0.405	710
Magnesium carbonate (magnesite)	$MgCO_3$	84.3	MgO	40.3	0.478	350*
Manganese carbonate	$MnCO_3$	114.9	MnO	70.9	0.615	300*
Manganese dioxide	MnO_2	86.9	MnO	70.9	0.816	1080*
Nepheline syenite	$K_2O \cdot 3Na_2O$ $4Al_2O_3 \cdot 8SiO_2$	1169.0	K_2O Na_2O Al_2O_3 SiO_2	94.2 62.0 101.9 60.1	0.081 0.159 0.349 0.411	1200
Nickel oxide	NiO	74.7	NiO	74.7	1	1990
Potash	K_2O	94.2	K_2O	94.2	1	700

Material	Formula	Molecular Weight	Oxides after firing	Oxide Weight	CF	MP (°C)
Potassium carbonate (pearl ash)	K_2CO_3	138.2	K_2O	94.2	0.682	896*
Silica (quartz)	SiO_2	60.1	SiO_2	60.1	1	1620
Silicon carbide	SiC	40.1	SiO_2	60.1	1.499	2700*
Soda	Na_2O	62.0	Na_2O	62.0	1	900
Sodium carbonate (soda crystals)	$Na_2CO_3 \cdot 1OH_2O$ (natron)	286.2	Na_2O	62.0	0.217	860*
Soda ash	Na_2CO_3	106	Na_2O	62.0	0.585	860*
Sodium chloride	NaCl	58.5	Na_2O	62.0	0.530	800
Sodium silicate (water glass)	Na_2SiO_3	122.1	Na_2O SiO_2	62.0 60.1	0.508 0.492	1088
Talc	$3MgO \cdot 4SiO_2 \cdot H_2O$	379.3	MgO SiO_2	40.3 60.1	0.319 0.634	900*
Tin oxide	SnO_2	150.7	SnO_2	150.7	1	1127
Titanium oxide	TiO_2	79.9	TiO_2	79.9	1	1600
Wollastonite	$CaSiO_3$	116.2	CaO SiO_2	56.1 60.1	0.483 0.517	1540
Zinkc oxide	ZnO	81.4	ZnO	81.4	1	1975
Zirconium silicate (zircon)	$ZrSiO_4$	183.3	ZrO_2 SiO_2	123.2 60.1	0.672 0.328	2550
Zirconium oxide	ZrO_2	123.2	ZrO_2	123.2	1	2700

Chemical Analysis of Glaze Materials

Ashes[1]

	SiO_2	Al_2O_3	Fe_2O_3	CaO	MgO	K_2O	Na_2O	P_2O_5	SO_3
Apple	2.7	–	–	70.9	5.5	11.8	1.9	4.5	2.7
Birch	11.5	–	1.3	29.6	14.3	22.6	9.0	7.9	2.0
Beech	3.9	–	0.6	42.0	8.2	24.3	8.3	6.2	2.1
Bone ash	2.9	–	0.3	52.4	1.3	–	–	40.9	–
Grass, washed	76.0	5.4	1.1	6.1	3.7	2.5	0.2	1.6	0.2
Oak	15.3	0.1	2.4	30.0	12.0	14.0	9.1	13.1	2.6
Oat straw	46.7	–	1.1	7.0	3.9	28.9	3.3	4.6	3.2
Poplar	3.5	0.2	1.2	16.5	10.2	54.2	2.3	10.2	1.3
Rice straw	83.2	1.4	0.4	3.7	1.4	2.5	0.4	0.7	0.4
Rice husk	96.0	1.0	0.04	0.2	0.5	0.2	0.9	0.02	–
Wheat straw	66.2	–	–	6.1	5.2	11.5	2.8	5.4	2.8

Rocks[2]

	SiO_2	Al_2O_3	Fe_2O_3	TiO_2	CaO	MgO	K_2O	Na_2O	L.o.I.
Basalt	49.1	15.7	5.4	1.3	8.9	6.2	1.5	3.1	1.6
Bauxite, India	6.8	55.0	4.5	7.8	–	–	–	–	23.8
Bentonite, India	54.5	18.7	3.7	0.2	2.3	2.6	1.3	2.1	7.6
Dolomite, Burma	2.6	0.7	0.3	–	30.8	19.8	–	–	45.4
Feldspar, Burma	65.2	19.1	0.1	–	–	–	11.9	3.5	0.2
Feldspar, Tanzania	67.0	19.7	0.1	0.1	1.25	0.1	2.85	3.21	–
Feldspar, India	64.0	18.8	0.4	–	0.6	–	12.0	3.1	0.2
Granite	70.2	14.5	1.6	0.4	1.9	0.9	4.11	3.5	–
Ilmenite	–	FeO:47.3	–	52.7	–	–	–	–	–
Kaolin, India, Bihar	47.5	38.7	0.6	0.9	0.2	0.1	0.1	0.1	12.4
Kaolin, Tanzania	48.0	35.5	1.6	0.7	0.1	0.1	0.2	0.2	13.2
Kyanite, Kenya	37.7	59.5	0.7	1.5	–	–	0.2	0.3	–
Lepidolite, U.S.A.	50.2	22.8	0.3	–	0.4	0.2	11.3	1.1	LiO_2:4.9
Limestone, Assam	7.4	3.3	3.2	–	45.6	0.9	–	–	38.8
Limestone, Tanzania	13.0	0.2	0.1	–	32.3	16.5	0.1	0.4	37.3
Magnesite, Greece	5.0	0.85	0.8	–	1.0	43.5	–	–	48.2
Nepheline syenite	50.2	30.4	0.1	–	0.7	–	3.4	12.9	0.4
Pegmatite	80.2	11.4	0.2	0.2	0.1	–	6.1	0.9	1.1
Petalite	77.7	17.8	–	–	–	–	0.2	0.3	LiO_2:4.2
Rhyolite	72.8	13.5	1.4	0.3	1.2	0.4	4.5	3.4	1.4
Talc, China	62.4	0.2	0.1	–	–	32.1	0.8	0.2	4.4
Volcanic ash	55.3	21.9	2.7	0.3	1.9	0.4	6.2	5.1	4.2
Zircon, Australia	33.0	0.1	0.1	0.1	–	–	–	–	ZrO_2:66.9
Zircon, India	25.9	4.0	0.2	–	0.1	0.6	0.1	0.5	ZrO_2:53.7

[1] W. Matthes: *Keramische Glasuren*, Verlag Müller, Darmstadt 1985.

[2] Sources: J. Singer & S.S. Singer. Central Glass & Ceramics Research Institute, India. Analysis carried out by ceramics projects in Africa and Asia sponsored by DANIDA, GTZ, NORAD.

Table of Standard Sieves[1]

Germany		Britain		United States		France
DIN		B.S.J.		U.S. standard		AFNOR
Openings mm	mesh/cm	Openings mm	mesh/in.	Openings mm	mesh/in.	Openings mm
6.000						
				5.613	$3\frac{1}{2}$	
5.000						5.000
				4.76	4	
4.000				4.00	5	4.000
		3.353	5	3.36	6	
3.000						3.150
		2.812	6	2.83	7	
2.500		2.411	7	2.38	8	2.500
2.000		2.057	8	2.00	10	2.000
1.500	4	1.676	10	1.68	12	1.600
		1.402	12	1.41	14	
1.200	5	1.204	14	1.19	16	1.250
1.000	6	1.003	16	1.00	18	1.000
		0.853	18	0.84	20	0.800
0.750	8	0.699	22	0.71	25	0.630
0.600	10	0.599	25	0.59	30	
0.500	12	0.500	30	0.50	35	0.500
0.430	14	0.422	36	0.42	40	0.400
0.400	16					
0.340	18	0.353	44	0.35	45	
0.300	20	0.295	52	0.297	50	0.315
0.250	24	0.251	60	0.250	60	0.250
0.200	30	0.211	72	0.210	70	0.200
0.177	35	0.178	85	0.177	80	
0.150	40	0.152	100	0.149	100	0.160
0.120	50	0.124	120	0.125	120	0.125
0.100	60	0.104	150	0.105	140	0.100
0.090	70	0.089	170	0.088	170	
0.075	80	0.076	200	0.074	200	0.080
0.060	100	0.066	240	0.062	230	0.063
0.053	110	0.053	300	0.053	270	0.050
0.040	130			0.037	400	0.040

Mesh means the number of threads per linear cm or inch of sieve cloth. Openings indicate the distance in mm between two threads.

[1] After F. and S.S. Singer, Industrial Ceramics.

Table of Seger Cone Formulas

Cone no.	Temperature °C	Formula		
022	600	Na_2O 0.5		SiO_2 2
		PbO 0.5		B_2O_3 1
021	650	Na_2O 0.5	Al_2O_3 0.02	SiO_2 1.04
		CaO 0.25		B_2O_3 1
020	670	Na_2O 0.5	Al_2O_3 0.04	SiO_2 1.08
		CaO 0.25		B_2O_3 1
		MgO 0.25		
019	690	Na_2O 0.5	Al_2O_3 0.08	SiO_2 1.16
		CaO 0.25		B_2O_3 1
018	710	Na_2O 0.5	Al_2O_3 0.13	SiO_2 1.26
		CaO 0.25		B_2O_3 1
		MgO 0.25		
017	730	Na_2O 0.5	Al_2O_3 0.2	SiO_2 1.4
		CaO 0.25		B_2O_3 1
		MgO 0.25		
016	750	Na_2O 0.5	Al_2O_3 0.31	SiO_2 1.61
		CaO 0.25		B_2O_3 1
		MgO 0.25		
015a	790	Na_2O 0.432	Al_2O_3 0.34	SiO_2 2.06
		CaO 0.432		B_2O_3 1
		MgO 0.136		
014a	815	Na_2O 0.385	Al_2O_3 0.34	SiO_2 1.92
		CaO 0.385		B_2O_3 0.77
		MgO 0.230		
013a	835	Na_2O 0.343	Al_2O_3 0.34	SiO_2 1.78
		CaO 0.343		B_2O_3 0.68
		MgO 0.314		
012a	855	Na_2O 0.345	Al_2O_3 0.365	SiO_2 2.04
		CaO 0.341		B_2O_3 0.68
		MgO 0.314		
011a	880	Na_2O 0.349	Al_2O_3 0.4	SiO_2 2.38
		CaO 0.340		B_2O_3 0.68
		MgO 0.311		
010a	900	Na_2O 0.338	Al_2O_3 0.423	SiO_2 2.626
		K_2O 0.011		B_2O_3 0.675
		CaO 0.338		
		MgO 0.313		
09a	920	Na_2O 0.336	Al_2O_3 0.468	SiO_2 3.087
		K_2O 0.018		B_2O_3 0.671
		CaO 0.335		
		MgO 0.311		
08a	940	Na_2O 0.279	Al_2O_3 0.543	SiO_2 2.691
		K_2O 0.038		B_2O_3 0.559
		CaO 0.369		
		MgO 0.314		

Cone no.	Temperatur °C	Formula		
07a	960	Na_2O 0.261 K_2O 0.055 CaO 0.391 MgO 0.293	Al_2O_3 0.554	SiO_2 2.984 B_2O_3 0.521
06a	980	Na_2O 0.247 K_2O 0.069 CaO 0.407 MgO 0.277	Al_2O_3 0.561	SiO_2 3.197 B_2O_3 0.493
05a	1000	Na_2O 0.229 K_2O 0.086 CaO 0.428 MgO 0.257	Al_2O_3 0.571	SiO_2 3.467 B_2O_3 0.457
04a	1020	Na_2O 0.204 K_2O 0.109 CaO 0.458 MgO 0.229	Al_2O_3 0.586	SiO_2 3.860 B_2O_3 0.407
03a	1040	Na_2O 0.182 K_2O 0.130 CaO 0.484 MgO 0.204	Al_2O_3 0.598	SiO_2 4.199 B_2O_3 0.363
02a	1060	Na_2O 0.157 K_2O 0.153 CaO 0.513 MgO 0.177	Al_2O_3 0.611	SiO_2 4.572 B_2O_3 0.314
01a	1080	Na_2O 0.134 K_2O 0.174 CaO 0.541 MgO 0.151	Al_2O_3 0.625	SiO_2 4.931 B_2O_3 0.268
1a	1100	Na_2O 0.109 K_2O 0.198 CaO 0.571 MgO 0.122	Al_2O_3 0.639	SiO_2 5.320 B_2O_3 0.217
2a	1120	Na_2O 0.085 K_2O 0.220 CaO 0.599 MgO 0.096	Al_2O_3 0.652	SiO_2 5.687 B_2O_3 0.170
3a	1140	Na_2O 0.059 K_2O 0.244 CaO 0.630 MgO 0.067	Al_2O_3 0.667	SiO_2 6.083 B_2O_3 0.170
4a	1160	Na_2O 0.043 K_2O 0.260 CaO 0.649 MgO 0.048	Al_2O_3 0.676	SiO_2 6.399 B_2O_3 0.119
5a	1180	Na_2O 0.028 K_2O 0.274 CaO 0.666 MgO 0.032	Al_2O_3 0.684	SiO_2 6.565 B_2O_3 0.056

Cone no.	Temperatur °C	Formula					
6a	1200	Na_2O	0.013	Al_2O_3	0.693	SiO_2	6.801
		K_2O	0.288			B_2O_3	0.020
		CaO	0.685				
		MgO	0.014				
7	1230	K_2O	0.3	Al_2O_3	0.7	SiO_2	7
		CaO	0.7				
8	1250	K_2O	0.3	Al_2O_3	0.8	SiO_2	8
		CaO	0.7				
9	1280	K_2O	0.3	Al_2O_3	0.9	SiO_2	9
		CaO	0.7				
10	1300	K_2O	0.3	Al_2O_3	1.0	SiO_2	10
		CaO	0.7				
11	1320	K_2O	0.3	Al_2O_3	1.2	SiO_2	12
		CaO	0.7				
12	1350	K_2O	0.3	Al_2O_3	1.4	SiO_2	14
		CaO	0.7				
13	1380	K_2O	0.3	Al_2O_3	1.6	SiO_2	16
		CaO	0.7				
14	1410	K_2O	0.3	Al_2O_3	1.8	SiO_2	18
		CaO	0.7				
15	1435	K_2O	0.3	Al_2O_3	2.1	SiO_2	21
		CaO	0.7				
16	1460	K_2O	0.3	Al_2O_3	2.4	SiO_2	24
		CaO	0.7				
17	1480	K_2O	0.3	Al_2O_3	2.7	SiO_2	27
		CaO	0.7				
18	1500	K_2O	0.3	Al_2O_3	3.1	SiO_2	31
		CaO	0.7				

Table of Seger Cones

°C	°F	German, Staffordshire, French Cone No.	New "H" series Staffordshire Cone No.	°C	°F	German, Staffordshire, French Cone No.	New "H" series Staffordshire Cone No.
600	1112	022	H 022	1100	2012	1 a	H 1
625	1157		H 022 A	1110	2030		H 1 A
650	1202	021	H 021	1120	2048	2 a	H 2
670	1238	020	H 020	1130	2066		H 2 A
690	1274	019	H 019	1140	2084	3 a	H 3
710	1310	018	H 018	1150	2102		H 3 A
730	1346	017	H 017	1160	2120	4 a	H 4
750	1382	016	H 016	1170	2138		H 4 A
790	1454	015	H 015	1180	2156	5 a	H 5
815	1499	014 a	H 014	1190	2174		H 5 A
835	1535	013 a	H 013	1200	2192	6 a	H 6
855	1571	012 a	H 012	1215	2219		H 6 A
880	1616	011 a	H 011	1230	2246	7	H 7
900	1652	010 a	H 010	1240	2264		H 7 A
920	1688	09 a	H 09	1250	2282	8	H 8
940	1724	08 a	H 08	1260	2300		H 8 A
960	1760	07 a	H 07	1270	2318		H 8 B
970	1778		H 07 A	1280	2336	9	H 9
980	1796	06 a	H 06	1290	2354		H 9 A
990	1814		H 06 A	1300	2372	10	H 10
1000	1832	05 a	H 05	1310	2390		H 10 A
1010	1850		H 05 A	1320	2408	11	H 11
1020	1868	04 a	H 04	1350	2462	12	H 12
1030	1886		H 04 A	1380	2516	13	H 13
1040	1904	03 a	H 03	1410	2570	14	H 14
1050	1922		H 03 A	1435	2615	15	H 15
1060	1940	02 a	H 02	1460	2660	16	H 16
1070	1958		H 02 A	1480	2696	17	H 17
1080	1976	01 a	H 01	1500	2732	18	H 18
1090	1994		H 01 A	1520	2768	19	H 19

Table of Orton Cones

(United States, Ohio, The E. Orton Jr. Ceramic Foundation)

Bending temperatures of large cones when heated at 150°/hour

°C	°F	Cone No.	°C	°F	Cone No.
600	1112	022	1120	2048	02
614	1137	021	1137	2079	01
635	1175	020	1154	2109	1
683	1261	019	1162	2124	2
717	1323	018	1168	2134	3
747	1377	017	1186	2167	4
792	1458	016	1196	2185	5
804	1479	015	1222	2232	6
838	1540	014	1240	2264	7
852	1566	013	1263	2305	8
884	1623	012	1280	2336	9
894	1641	011	1305	2381	10
894	1641	010	1315	2399	11
923	1693	09	1326	2419	12
955	1751	08	1346	2455	13
984	1803	07	1366	2491	14
999	1830	06	1431	2608	15
1046	1915	05	1473	2683	16
1060	1940	04	1485	1705	17
1101	2014	03	1506	2743	18

Note:
The temperatures indicated in these cone tables may not be the same as when the cones bend in the individual potter's kiln. Cones are not used for measuring temperatures but for indicating the condition of clay and glazes.

Conversion Table for Pint Weights

oz/pt UK	oz/pt US	S.G.	°TW
22	18.3	1.10	20
22.8	19	1.14	28
23	19.2	1.15	30
24	20	1.20	40
25	20.8	1.25	50
25.2	21	1.26	52
26	21.7	1.30	60
26.4	22	1.32	64
27	22.5	1.35	70
27.6	23	1.38	76
28	23.3	1.40	80
28.8	24	1.44	88
29	24.2	1.45	90
30	25	1.50	100
31	25.8	1.55	110
31.2	26	1.56	112
32	26.7	1.60	120
32.4	27	1.62	124
33	27.5	1.65	130
33.6	28	1.68	136
34	28.3	1.70	140
34.8	29	1.74	148
35	29.2	1.75	150
36	30	1.80	160
37	30.8	1.85	170
37.2	31	1.86	171
38	31.6	1.89	179

DENSITY

Specific gravity (SG) of a material, a mixture of materials or a clay slip is expressed as how many times it is heavier than the same amount of water, i.e. how many kg per 1 liter volume or gram per cm^3. Density is the weight per volume unit and in the metric system this equals specific gravity (g/cc or kg/l) but in many countries slip densities are still measured in ounces per pint.

The density of a clay slip is found by weighing 1 liter of the slip. If it weighs 1.6 kg the slip has a density of 1.6.

DRY CONTENT OF A LIQUID

Brogniart's Formula.

It is often useful to know the dry weight of materials in liquid clay slips or glazes. First find the weight of 1 liter of the liquid. The density (specific gravity, g/cm^3) of the dry material has to be known. For clay materials it is close to 2.5 . Density of glazes has to be calculated from the density of the materials in the glaze recipe.

$$\text{Dry weight in g} = \frac{(W - 1000) \times D}{D - 1}$$

W = weight in g of 1 liter liquid
D = density of dry material

TWADDELL SCALE

Clay and glaze suspensions have normally densities between 1.0 and 2.0. On hydrometers used for measuring glaze and slip densities the densities between 1.0 and 2.0 have been divided into 200 units. These units are called degrees Twaddell and the formula for calculating these is:

$$^{o}TW = (\text{density} - 1) \times 200$$

$$\text{Density} = \frac{^{o}TW}{200} + 1$$

Properties of fuels

Average properties of solid fuels[1]

		Wood	Peat	Lignite	Bituminous Coal	Charcoal
Moisture content as found	%	25–50	90	50	2	
Moisture content at firing	%	10–15	15–20	15	2	2
Volatile matters	%	80	65	50	30	10
fixed carbon	%	20	30	45	65	89
ash	%	trace	5	5	5	1
Chemical analysis:						
carbon, C	%	50.0	57.5	70.0	86.0	93.0
hydrogen, H	%	6.0	5.5	5.0	5.5	2.5
oxygen, O	%	43.0	35.0	23.0	6.0	3.0
nitrogen + sulphur		1.0	2.0	2.0	2.5	1.5
Calorific value: cal/g						
dry fuel	gross	4450	5000	6400	8600	8300
	net	4130	4710	6140	8310	8170
normal fuel	gross	3780	3800	5170	8000	8050
	net	3420	3460	4870	7720	7910

Properties of dry wood[2]

		Specific gravity	Ash %	cal/g
Hardwood:	ash	.74	.6	4450
	beech	.68	.6	4500
	oak	.83	.4	4360
softwood:	fir	.45	.3	4770
	pine	.48	.4	4820
	elm	.56	.5	4470

Note:
Heat or calorific value is measured in calories per gram of fuel. One calorie is the heat required to heat 1 gram of water 1 °C.
Gross calorific value is the heat that theoretically can be obtained, whereas net value is what is normally obtained when firing a kiln. Both values are included for comparison with other fuels.

[1] F. Singer and S.S. Singer: "Industrial Ceramics" p. 904
[2] Brame and King: "Fuel" p. 37

Properties of liquid fuels

	Waste oil	Heavy fuel oil	Medium fuel oil	Light fuel oil	Kerosene
Specific gravity	0.9–1	1.1–0.94	0.93–0.91	0.9–0.81	0.78
flash point °C	250	200	150	105	55
viscosity	very high	high	medium	low	very low
calorific value:					
cal/g gross	10300	10055	10130	10300	11100
net		9480	9536	9695	

Metric system

1 kilometre, km	=	1000 metres, m
1 m	=	100 centimetres, cm
1 cm	=	10 millimetres, mm
1 mm	=	1000 micron, μ

1 cubic metre, m^3	=	1000 litres, l
1 l	=	1000 cm^3 or ml
1 ml	=	1000 mm^3
1 ton	=	1000 kilograms, kg
1 kg	=	1000 grams, g
1 g	=	1000 milligrams, mg

to convert:	to:	multiply metric by:	to convert:	to:	multiply UK & US by:
length:					
m	feet	3.280	feet	m	.305
m	inches	39.370	inches	m	.025
cm	inches	.394	inches	cm	2.54
mm	inches	.039	inches	mm	25.400
area:					
hectare	acres	2.471	acres	hectare	.405
m^2	sq. feet	10.764	sq. feet	m^2	.093
cm^2	sq. inches	.155	sq. inches	cm^2	6.451
volumne:					
m^3	cu. feet	35.314	cm feet	m^3	0.0283
m^3	cord (wood)	.276	cord	m^3	3.625
cm^3 (cc)	cu. inches	.061	cu. inches	cm^3 (cc)	16.387
l	cu. inches	61.020	cu. inches	l	.016
l	U.K gallon	.219	U.K gallon	l	4.546
l	U.S gallon	.264	U.S gallon	l	3.785
weight:					
kilograms	pounds	2.205	pounds	kilograms	.453
grams	ounces	0.035	ounces	grams	28.349

Temperature Conversion Table

°C	0	10	20	30	40	50	60	70	80	90
					Degrees Fahrenheit					
0	32	50	68	86	104	122	140	158	176	194
100	212	230	248	266	284	302	320	338	356	374
200	392	410	428	446	464	482	500	518	536	554
300	572	590	608	626	644	662	680	698	716	734
400	752	770	788	806	824	842	860	878	896	914
500	932	950	968	986	1004	1022	1040	1058	1076	1094
600	1112	1130	1148	1166	1184	1202	1220	1238	1256	1274
700	1292	1310	1328	1346	1364	1382	1400	1418	1436	1454
800	1472	1490	1508	1526	1544	1562	1580	1598	1616	1634
900	1652	1670	1688	1706	1724	1742	1760	1778	1796	1814
1000	1832	1850	1868	1886	1904	1922	1949	1958	1976	1994
1100	2012	2030	2048	2066	2084	2102	2120	2138	2156	2174
1200	2192	2210	2228	2246	2264	2282	2300	2318	2336	2354
1300	2372	2390	2408	2426	2444	2462	2480	2498	2516	2534
1400	2552	2570	2588	2606	2624	2642	2660	2678	2696	2714
1500	2732	2750	2768	2786	2804	2822	2840	2858	2876	2894
1600	2912	2930	2948	2966	2984	3002	3020	3038	3056	3074
1700	3092	3110	3128	3146	3164	3182	3200	3218	3236	3254
1800	3272	3290	3308	3326	3344	3362	3380	3398	3416	3434
1900	3452	3470	3488	3506	3524	3542	3560	3578	3596	3614
2000	3632	3650	3668	3686	3704	3722	3740	3758	3776	3794

°C	1	2	3	4	5	6	7	8	9	10
°F	1.8	3.6	5.4	7.2	9.0	10.8	12.6	14.4	16.2	18.0

Conversion Formulas:

$$X\,°C = \frac{X \times 9}{5} + 32\,°F$$

$$Y\,°F = \frac{Y - 32}{9} \times 5\,°C$$

Example: 573 °C = 1058° + 5.4° = 1063.4 °F

500 is found in the left column and the 70 is found at the top. The equivalent of 570 is the crossing point and then the final digit is added.

BIBLIOGRAPHY

J.L.Artigas. Formulario y Practicas de Ceramica, Barcelona 1961.
Michael Cardew. Pioneer Pottery, New York 1971.
James Chappel. The Potter's Complete Book of Clay and Glazes, London 1977.
Emmanuel Cooper and Derek Royle. Glazes for The Studio Potter, London 1984.
Ivan Englund. Rock Glazes, Walcha 1983.
Harry Fraser. Ceramic Faults and their Remedies, London 1986.
Harry Fraser. Glazes for the Craft Potter, London 1984.
John Gibson. Pottery Decoration, London 1987.
D. Green. A Handbook of Pottery Glazes, London 1978.
R. Griffiths and C. Radford. Calculations in Ceramics, London 1965.
Frank Hamer. The Potter's Dictionary of Materials and Techniques, London 1975.
W.G. Lawrence. Ceramic Science For The Potter, Radnor 1982.
Bernard Leach. A Potter's Book, London 1976.
Werner Lehnhäuser. Glasuren und ihre Farben, Düsseldorf 1978.
Wolf E. Matthes. Keramische Glasuren, Darmstadt 1985.
David C. Maynard. Ceramic Glazes, London 1980.
S.K. Mirmira. Indian Pottery, New Delhi 1987.
Glenn C. Nelson. Ceramics, New York 1960.
Henrik Norsker. Clay Materials for the Self-Reliant Potter, Eschborn 1990.
Daniel Rhodes. Clay and Glazes for the Potter, New York 1966.
Thomas Shafer. Pottery Decoration, New York 1976.
Kenneth Shaw. Ceramic Glazes, London 1971.
F.Singer and S.S. Singer. Industrial Ceramics, London 1963.
Stefanov/Batschwarov. Keramik-Glasuren, Ceramic Glazes, Berlin 1988.
Brian Sutherland. Glazes from Natural Sources, London 1987.
J.R. Taylor and A.C. Bull. Ceramics Glaze Technology, Oxford 1986.
The Institute of Ceramics. Health & Safety in Ceramics, Oxford 1986.
Jack Troy. Salt-Glazed Ceramics, New York 1977.

CERAMICS JOURNALS

Ceramics Monthly, Box 12448, Columbus, Ohio, U.S.A.
Interceram, Verlag Schmid, P.O. Box 6609, 7800 Freiburg, Germany.
Tile & Brick International, Verlag Schmid.
New Zealand Potter, P.O. Box 12-162, Wellington, New Zealand.
Ceramic Review, 21 Carnaby Street, London, U.K.
Studio Potter, Box 70, Goffstown, N.H., U.S.A.

Index

GPSR Compliance

The European Union's (EU) General Product Safety Regulation (GPSR) is a set of rules that requires consumer products to be safe and our obligations to ensure this.

If you have any concerns about our products, you can contact us on ProductSafety@springernature.com

In case Publisher is established outside the EU, the EU authorized representative is:

Springer Nature Customer Service Center GmbH
Europaplatz 3
69115 Heidelberg, Germany

Batch number: 10730990

Printed by Printforce, the Netherlands